GLOBAL AGRICULTURE DEVELOPMENTS

GLOBAL AGRICULTURAL INDUSTRIES

RESEARCH SPENDING TRENDS AND CHANGING MARKET STRUCTURES

GLOBAL AGRICULTURE DEVELOPMENTS

Additional books in this series can be found on Nova's website
under the Series tab.

Additional E-books in this series can be found on Nova's website
under the E-book tab.

AGRICULTURE ISSUES AND POLICIES

Additional books in this series can be found on Nova's website
under the Series tab.

Additional E-books in this series can be found on Nova's website
under the E-book tab.

GLOBAL AGRICULTURE DEVELOPMENTS

GLOBAL AGRICULTURAL INDUSTRIES

RESEARCH SPENDING TRENDS AND CHANGING MARKET STRUCTURES

SHIRLEY O. KELLEY

AND

MIKE F. GRANT

EDITORS

nova publishers

Nova Science Publishers, Inc.

New York

For permission to use material from this book please contact us:
Telephone 631-231-7269; Fax 631-231-8175
Web Site: http://www.novapublishers.com

Library of Congress Cataloging-in-Publication Data

ISBN 978-1-62257-476-6

Published by Nova Science Publishers, Inc. † New York

CONTENTS

PREFACE

Growth in the productivity of the global food and agricultural system will be largely determined by today's investments in research and development (R&D). In recent decades, the private sector has become a major player in developing innovations for food and agriculture. Factors spurring private companies to invest in food and agricultural research include the emergence of biotechnology and other new scientific developments, the strengthening of intellectual property rights (IPR) over agricultural innovations, new regulatory requirements, the expansion of markets for improved agricultural inputs and food products, and rising consumer demand for more diverse foods. This book evaluates investment trends by for-profit companies in food manufacturing, biofuel, and agricultural input R&D and explores how these trends are affected by changes in market demand and industry structure.

Chapter 1 - Meeting growing global demand for food, fiber, and biofuel requires robust investment in agricultural research and development (R&D) from both public and private sectors. This study examines global R&D spending by private industry in seven agricultural input sectors, food manufacturing, and biofuel and describes the changing structure of these industries. In 2007 (the latest year for which comprehensive estimates are available), the private sector spent $19.7 billion on food and agricultural research (56 percent in food manufacturing and 44 percent in agricultural input sectors) and accounted for about half of total public and private spending on food and agricultural R&D in high-income countries. In R&D related to biofuel, annual private-sector investments are estimated to have reached $1.47 billion worldwide by 2009. Incentives to invest in R&D are influenced by market structure and other factors. Agricultural input industries have undergone significant structural change over the past two decades, with industry concentration on the rise. A relatively small number of large, multinational firms with global R&D and marketing networks account for most R&D in each input industry. Rising market concentration has not generally been associated with increased R&D investment as a percentage of industry sales.

Chapter 2 - By 2050, global agricultural demand is projected to grow by 70-100 percent due to population growth, energy demands, and higher incomes in developing countries. Meeting this demand from existing agricultural resources will require raising global agricultural total factor productivity (TFP) by a similar level. Maintaining the U.S. contribution to global food supply would also require a similar rise in U.S. agricultural TFP.

TFP growth in U.S. agriculture is predicated on long-term investments in public agricultural research and development (R&D). Productivity growth also springs from agricultural extension, farmer education, rural infrastructure, private agricultural R&D, and technology transfers, but the force of these factors is compounded by public agricultural research.

The rate of TFP growth (and therefore output growth) of U.S. agriculture has averaged about 1.5 percent annually over the past 50 years. Stagnant (inflation-adjusted) funding for public agricultural research since the 1980s may be causing agricultural TFP growth to slow down, although statistical analyses of productivity growth trends are inconclusive.

ERS simulations indicate that if U.S. public agricultural R&D spending remains constant (in nominal terms) until 2050, the annual rate of agricultural TFP growth will fall to under 0.75 percent and U.S. agricultural output will increase by only 40 percent by 2050. Under this scenario, raising output beyond this level would require bringing more land, labor, capital, materials, and other resources into production.

Additional public agricultural R&D spending would raise U.S. agricultural productivity and output growth. Raising R&D spending by 3.73 percent annually (offsetting the historical rate of inflation in research costs) would increase U.S. agricultural output by 73 percent by 2050. Raising R&D spending by 4.73 percent per year (1-percent annual growth in inflation-adjusted spending) would increase output by 83 percent by 2050.

In: Global Agricultural Industries
Editors: Shirley O. Kelley and Mike F. Grant

ISBN: 978-1-62257-476-6

Chapter 1

RESEARCH INVESTMENTS AND MARKET STRUCTURE IN THE FOOD PROCESSING, AGRICULTURAL INPUT, AND BIOFUEL INDUSTRIES WORLDWIDE*

Keith O. Fuglie, Paul W. Heisey, John L. King, Carl E. Pray, Kelly Day-Rubenstein, David Schimmelpfennig, Sun Ling Wang and Rupa Karmarkar-Deshmukh

ABSTRACT

Meeting growing global demand for food, fiber, and biofuel requires robust investment in agricultural research and development (R&D) from both public and private sectors. This study examines global R&D spending by private industry in seven agricultural input sectors, food manufacturing, and biofuel and describes the changing structure of these industries. In 2007 (the latest year for which comprehensive estimates are available), the private sector spent $19.7 billion on food and agricultural research (56 percent in food manufacturing and 44 percent in agricultural input sectors) and accounted for about half of total public and private spending on food and agricultural R&D in high-income countries. In R&D related to biofuel, annual private-sector investments are estimated to have reached $1.47 billion worldwide by 2009. Incentives to invest in R&D are influenced by market structure and other factors. Agricultural input industries have undergone significant structural change over the past two decades, with industry concentration on the rise. A relatively small number of large, multinational firms with global R&D and marketing networks account for most R&D in each input industry. Rising market concentration has not generally been associated with increased R&D investment as a percentage of industry sales.

* This is an edited, reformatted and augmented version of United States Department of Agriculture, Economic Research Report No. 130, dated December 2011.

SUMMARY

What Is the Issue?

Growth in the productivity of the global food and agricultural system will be largely determined by today's investments in research and development (R&D). In recent decades, the private sector has become a major player in developing innovations for food and agriculture. Factors spurring private companies to invest in food and agricultural research include the emergence of biotechnology and other new scientific developments, the strengthening of intellectual property rights (IPR) over agricultural innovations, new regulatory requirements, the expansion of markets for improved agricultural inputs and food products, and rising consumer demand for more diverse foods. More recently, rapid growth in the market for biofuel has pushed companies to expand their R&D investments in this area as well.

This report quantifies investment trends by for-profit companies in food manufacturing, biofuel, and agricultural input R&D and explores how these trends are affected by changes in market demand and industry structure. In particular, the report examines changes in the organization and structure of agricultural input industries (crop seed and biotechnology, crop protection chemicals, synthetic fertilizers, farm machinery, animal breeding and genetics, animal health, and animal nutrition) and whether increases in market concentration in these industries are associated with increases or decreases in the level and intensity of R&D investments.

For comparative purposes, we present some aggregate statistics on public-sector research spending for food and agriculture and ways in which these investments differ or complement R&D in the private sector. However, we do not delve much into the interactions between public and private R&D. For a detailed examination of the evolving role of the public and private sectors in agricultural R&D in the United States, see Fuglie and Schimmelfpennig (2000).

What Did the Study Find?

During 1994-2007 (the latest year for which estimates are available), annual private-sector food and agricultural R&D grew from $11.3 billion to $19.7 billion, or 4.3 percent per year (or, in constant 2006 dollars, from $14.6 billion to $19.2 billion, or 2.1 percent per year). In high-income countries, private-sector R&D spending appeared to be roughly equivalent to public-sector spending on food and agricultural R&D, although public R&D spending continues to be larger if only agricultural-related R&D is considered.

Growth in R&D investment was uneven across industries. The most rapid increase in R&D was in crop breeding/biotechnology. Significant growth in R&D spending also occurred in farm machinery and food manufacturing. However, real (inflation-adjusted) R&D spending declined for crop protection chemicals and animal nutrition.

Other key findings include the following (figures below are in current or nominal dollars, unadjusted for inflation):

- In 2010, global private-sector investments in R&D related to *agricultural inputs* reached $11.03 billion, an increase from $5.58 billion in 1994.
- In 2007, global private-sector investments in R&D related to *food manufacturing* reached $11.48 billion, an increase from $6.02 billion in 1994.
- In 2009, global private-sector investments in R&D related to *biofuel reached* $1.47 billion, with most growth in this area occurring since 2000.
- Generally, the largest four to eight firms in each sector accounted for about three-fourths of the R&D in that sector, with larger firms spending more than smaller firms on R&D as a percentage of product sales (with the exception of small biotechnology firms). Typically, the large firms are multinational operations with global R&D and marketing networks.
- In most of the agricultural input industries, market concentration increased during 1994-2009, with the highest levels observed in the animal breeding and crop seed sectors and the largest increase observed in the crop seed sector.
- Rising levels of market concentration were not associated with larger R&D investment in agricultural input sectors.
- The globalization of food and agricultural R&D may accelerate the rate of international technology transfer, reducing productivity differences across nations and regions.

How Was the Study Conducted?

We used a number of approaches to construct estimates of private R&D spending by sector. For research-intensive agricultural input industries, we built a database of agriculturally related research spending firm by firm over time, for all firms in the sector (including "legacy" firms, or firms that exited the industry during the period of study) that have or have had significant R&D expenditures. For large conglomerates, for which agriculture may be only one business segment, we separated agriculturally related R&D spending from R&D spending on nonagricultural business segments. We gathered this information by canvassing a broad set of material, including company annual reports and websites, reports by industry associations and consulting services, and personal interviews with company representatives. Altogether, we reviewed R&D information on more than 800 agricultural input companies worldwide. These firm-level data also enabled us to examine hypotheses regarding the relationship between industry structure and R&D spending: Do larger firms spend more (as a percentage of product sales) on R&D than smaller firms? Has the rising concentration of several agricultural input industries affected overall levels of R&D spending by that industry?

For agricultural input industries in which firms do not often report their research spending, we estimated agricultural R&D for the industry by taking a percentage of total agricultural input sales, with the percentages (or research intensities) derived from observations on R&D spending from a subset of firms and from previous surveys of the industry. For the food manufacturing industry, we relied on country-level estimates produced by the Organisation for Co-operation and Development, which covers primarily high-income countries.

With these sources, we developed a global time series of R&D expenditures by agricultural input industries from 1994 to 2010, the food industry from 1990 to 2007, and for biofuel in 2009. We examined how trends in R&D spending were associated with changes in market demand and industry structure and reviewed the evidence on the factors causing structural changes in agricultural input industries.

Research Investments and Market Structure in the Food Processing, Agricultural Input, and Biofuel Industries Worldwide: Synthesis of Results

Objectives and Methods of Study

Over the past several decades, the private sector has become a major player in developing new innovations for food and agriculture. The emergence of biotechnology and other new scientific developments, the strengthening of intellectual property rights (IPR) over agricultural innovations, the global expansion of markets for improved agricultural inputs and food products, and consumer demands for more diverse kinds of food products are some of the key factors driving private companies to invest in food and agricultural research. More recently, rapid growth in the market for biofuel has spurred a diverse set of firms to expand their R&D investments in this area as well. This report seeks to quantify investment trends by for-profit companies in agricultural, food, and biofuel R&D and explore how changing market demand, industrial structure, and public policy may be affecting these trends. In addition, the report examines the role of government subsidies in stimulating private R&D in the biofuel sector.

Existing information on private spending on food and agricultural research is fragmentary. James (1997) and Alston et al. (2010) are among the few studies that have attempted to provide estimates of such expenditures on a global scale. Based on findings from both studies, private R&D expenditures from the mid-1990s to 2000 are estimated at $13 billion per year, or about two-thirds of total public sector spending for agricultural R&D (about $20 billion per year globally) over the period. These estimates account for R&D by the food manufacturing sector and the agricultural input industries, but the studies did not break down these amounts by sector. Moreover, they provide limited detail (and quite different estimates) about the country-specific locations of private-sector R&D, with James estimating that about 85 percent was conducted in high-income countries and Alston et al. putting the share at 95 percent (in contrast, about 60 percent of public agricultural R&D is conducted in high-income countries, according to Alston et al.).

Other studies have provided more detailed information on private-sector expenditures on food and agricultural R&D at the country level. Klotz et al. (1995) develop comprehensive estimates of private R&D by the food sector and for major agricultural input industries in the United States between 1960 and 1992. Pray and Fuglie (2001) survey private companies in seven Asian countries about their agricultural R&D investments in the mid-1990s, and Echeverria et al. (1996) summarize available information for eight Latin American countries from around the same period.

Some estimates of R&D in specific industries, such as the agricultural chemical, crop seed, and veterinary pharmaceutical industries, are provided by industry groups through surveys of their member companies or consulting services. This information, however, may cover only a portion of an industry and may not be in the public domain.

Finally, a number of studies have examined publicly available data on a range of indicators of private R&D effort, such as number of agricultural patents, plant variety protection certifications, and biotechnology field trials issued or undertaken. For example, Huffman and Evenson (2006) make extensive use of historical patent data to investigate technology flows from manufacturing sectors to agriculture in the United States. The main conceptual difference between these indicators and R&D expenditures is that the indicators reflect *outputs* from the R&D process whereas expenditures measure R&D *inputs*. It is expected that the two would be significantly correlated but with a time lag. Some of the main findings from studies assessing agricultural R&D indicators are summarized in Pray et al. (2007). In this study, we extend some of the work on R&D output indicators in the chapters on crop seed and agricultural chemicals.

This study provides new, detailed information on R&D spending by private industry for the food processing and biofuel sectors and for seven agricultural input sectors (crop seed and biotechnology, crop protection chemicals, synthetic fertilizers, farm machinery, animal health, animal breeding and genetics, and animal nutrition). For the food processing and agricultural input sectors, we report trends in private R&D spending over time. For the newly emerging biofuel sector, our estimates cover only one year, 2009. We also examine the location of private-sector R&D, but the multinational nature of many of the leading companies conducting food and agricultural R&D makes it difficult to do so. For agricultural input sectors, we can estimate total R&D for companies based in a particular country, but this estimate includes R&D by those same companies conducted in other countries and excludes R&D by foreign companies in that country. We discuss the globalization of private-sector R&D in terms of the growing international trade in agricultural inputs and how companies locate their R&D facilities to serve global markets.

The study also examines the changing structure of agricultural input industries. Several of these industries have undergone significant consolidation over the past couple of decades, with many firms exiting, merging, or being acquired by other firms. We discuss factors causing these changes and, for the agricultural input industries that do the most research, we quantify the change in concentration at the global level. Higher levels of concentration may impart greater market power to the largest firms in the industry. If this market power is exercised to raise premiums on firms' proprietary technology, it could encourage these firms to invest more in R&D. We examine whether market concentration is correlated with the share of industry revenues that is invested in R&D. We do not, however, conduct any formal tests of competitive performance in these markets.

To construct estimates of private R&D spending by sector, we use a number of approaches. For research-intensive agricultural input industries, we build a database of agriculturally related research spending firm by firm (both publicly traded and privately held) over time, for all firms in the sector that have (or have had) significant R&D expenditures. For large conglomerates, in which agriculture may be only one line of business among many, we separate agriculturally related R&D spending from other R&D spending. We gather this information primarily from firms' annual financial reports and supplement it with information from industry associations, consulting services, and personal interviews with company representatives. These firm-level data also enable us to address questions on the relationship between industry structure and R&D spending:

- Do larger firms spend more (as a percentage of product sales) on R&D than smaller firms?
- Has the rising concentration of several agricultural input industries affected overall levels of R&D spending by these industries?

For agricultural input industries in which member firms do not conduct much research, firm-level data on R&D spending is often reported for only a subset of the major companies in the industry. Our estimates of agricultural R&D for such industries reflect a share of total agricultural input sales (or research intensities) derived from observations on R&D spending from a sample of firms in the respective industries.

For the food industry, we rely primarily on country-level estimates provided by the Analytical Business Enterprise Research and Development (ANBERD) database produced by the Organisation for Economic Co-operation and Development (OECD). This database covers most high-income countries and a few developing countries.

For biofuel, we examine R&D spending across a number of sectors that are developing technologies for both biofuel feedstocks and biofuel manufacturing. Because this is a relatively young industry, we derive an estimate of private R&D for only one year, 2009.

From these sources, we are able to develop a global time series of R&D expenditures by agricultural input industries from 1994 to 2010, food processing industries from 1990 to 2007, and the biofuel industry for 2009. Significant overlap or duplication occurs between R&D reported in the food and agricultural sector and the different segments of the biofuel market chain (i.e., some seed industry R&D is directed toward biofuel feedstocks and is counted as R&D in both sectors); therefore, to avoid double counting, we report biofuel R&D estimates separately from the estimate for total private-sector food and agricultural R&D.

Having assembled data on trends and levels of private food manufacturing and agricultural input R&D spending, we examine several factors that may be influencing these trends. First, we look at market demand. Large and growing markets for agricultural inputs or new food products can be expected to attract more R&D from private firms seeking to meet these needs. Second, we examine industry structure. Mergers and acquisitions have affected many agricultural input industries examined, with the result that fewer firms account for a growing share of the market over time. This development could influence incentives for private R&D positively, negatively, or not at all. The classic Schumpeterian view is that larger firms invest a greater portion of their revenues in R&D than smaller firms. However, in a detailed study of U.S. manufacturing industries, Cohen et al. (1987) do not find empirical support for this hypothesis. Regarding concentration, Levin et al. (1985) report a general tendency for R&D intensity to first increase and then decrease as industry concentration rises, but the authors note that the differences across industries can be much larger than changes within an industry.

Finally, we discuss the effects of changes in policies and technology opportunity, namely, the influence of developments in biotechnology on structure and R&D in the research-intensive agricultural input industries. Policies toward intellectual property rights (what is considered patentable) and the regulation of new technology introductions may have significant effects on how much and what kind of R&D is undertaken by the private sector, and what kinds of firms can successfully navigate these policies.

Private-Sector R&D Investment in Agriculture, Food, and Biofuel

R&D Spending over Time

Table 1.1 shows trends in private-sector R&D spending in various agricultural input sectors and the food manufacturing industry in both nominal and constant (inflation-adjusted) dollars. In constant 2006 U.S. dollars, total food and agricultural R&D expenditures in the private sector increased from $14.59 billion in 1994 to $19.18 billion in 2007, or at an average annual rate of 2.1 percent. R&D expenditures in food manufacturing rose faster than those in agricultural input industries, and by 2007, food manufacturing accounted for about 58 percent of the overall annual total. Food manufacturing has relatively low research intensity (R&D as a percentage of sales), but the overall size of the market is very large. R&D in the industry appears to be directed mostly toward new product development. Food sector R&D that is directly relevant to agriculture, such as R&D on animal feed manufacturing, is also included in our estimate of R&D in agricultural input industries (but not double-counted in the total for food and agriculture). Among agricultural input industries, most of the increase in R&D spending between 1994 and 2010 occurred in the crop input industries, with R&D spending in the animal-related sectors as a whole remaining essentially flat in real (inflation-adjusted) dollars. Across sectors, the most rapid growth in agricultural R&D over 1994-2010 was for crop seed and biotechnology, where annual R&D spending increased from about $1.5 billion in the mid-1990s to nearly $3.5 billion in 2010 (constant 2006 U.S. dollars). Real R&D spending declined for crop protection chemicals and animal nutrition.

Comparative statistics for government spending on agricultural research are only available for 2000 (Beintema and Stads, 2008; Alston et al., 2010, table 6-1). Beintema and Stads (2008) estimate that total global public-sector agricultural research in 2000 was $16.3 billion in U.S. dollars and $20.8 billion in purchasing-power-parity (PPP) dollars.[1]

The private sector appears to account for between 39 and 45 percent of the total global investment in food and agricultural R&D worldwide, depending on whether comparisons are made using market or PPP exchange rates, and about half of the total in high-income countries (table 1.2). For high-income countries, Beintema and Stads estimate total public agricultural R&D in 2000 was $12.3 billion in U.S. dollars and $11.8 billion in PPP dollars, respectively. Of our estimated total of U.S. $13.1 billion (PPP $13.2 billion) in private food and agricultural R&D in 2000, U.S. $12.2 billion (PPP $11.8 billion) was attributed to companies based in high-income countries.

Although none of the global estimates of public research spending break down this investment into food and agricultural sectors, the U.S. data may be illustrative, at least for high-income countries. According to USDA's Inventory of Agricultural Research, in 2000, about 60 percent of total public agricultural R&D was allocated to research related to plant and animal systems, 15 percent went to food and human nutrition, 18 percent went to environmental issues, and the remaining 7 percent was spread across other topics not directly related to food or farm production.[2]

Table 1.1. Private research and development (R&D) expenditures for food and agriculture worldwide

	Crop protection chemicals	Crop seed & biotech.	Farm machinery	Fertilizer	Food animal health[1]	Animal breeding & genetics[2]	Animal nutrition	Total crop inputs	Total animal inputs	Total agricultural inputs	Food manufacturing	Total food & agricultural inputs[3]
					Millions of nominal U.S. dollars							
1994	2,296	1,130	920	61	664	196	314	4,407	1,173	5,579	6,016	11,282
1995	2,390	1,213	987	80	778	203	332	4,670	1,313	5,983	6,876	12,528
1996	2,523	1,322	1,110	84	767	210	373	5,039	1,350	6,389	6,468	12,483
1997	2,635	1,522	1,127	64	749	217	345	5,349	1,311	6,660	6,399	12,714
1998	2,636	1,721	1,164	56	720	225	324	5,577	1,269	6,846	6,417	12,939
1999	2,581	1,788	1,079	49	670	232	320	5,496	1,223	6,719	6,490	12,889
2000	2,352	2,055	1,197	56	655	240	329	5,659	1,224	6,883	6,516	13,071
2001	2,263	2,015	1,149	53	592	249	334	5,480	1,175	6,655	6,755	13,075
2002	2,076	1,976	1,136	56	590	258	345	5,245	1,193	6,438	7,203	13,295
2003	2,458	2,064	1,190	74	663	267	360	5,787	1,290	7,076	8,756	15,472
2004	2,628	2,180	1,275	97	712	276	377	6,181	1,365	7,545	9,620	16,789
2005	2,678	2,254	1,369	119	757	285	375	6,420	1,417	7,837	10,531	17,993
2006	2,633	2,374	1,470	99	794	295	375	6,575	1,465	8,040	10,899	18,564
2007	2,754	2,615	1,665	104	816	306	389	7,138	1,511	8,649	11,480	19,741
2008	3,012	3,093	2,003	96	960	316	400	8,205	1,677	9,882	n.a.	n.a.
2009	2,987	3,342	2,310	100	930	327	405	8,739	1,663	10,402	n.a.	n.a.
2010	3,116	3,726	2,394	100	941	339	410	9,335	1,690	11,026	n.a.	n.a.
					Millions of constant 2006 U.S. dollars							
1994	2,968	1,462	1,189	79	858	253	405	5,697	1,516	7,214	7,778	14,587
1995	3,028	1,536	1,250	101	986	257	421	5,915	1,663	7,578	8,709	15,866
1996	3,136	1,643	1,380	104	953	261	464	6,263	1,678	7,941	8,039	15,516
1997	3,218	1,859	1,377	79	915	265	421	6,533	1,601	8,134	7,815	15,528
1998	3,183	2,078	1,406	67	870	271	391	6,735	1,533	8,268	7,749	15,626
1999	3,071	2,127	1,284	58	798	277	381	6,541	1,455	7,996	7,724	15,339
2000	2,739	2,394	1,395	65	763	280	383	6,592	1,425	8,018	7,590	15,225
2001	2,577	2,295	1,309	61	674	283	381	6,242	1,338	7,580	7,694	14,894
2002	2,328	2,215	1,274	63	662	289	387	5,880	1,337	7,217	8,075	14,905
2003	2,697	2,265	1,306	81	727	292	396	6,350	1,415	7,765	9,609	16,978
2004	2,805	2,326	1,361	104	760	294	402	6,595	1,456	8,052	10,265	17,915
2005	2,765	2,328	1,414	123	781	295	387	6,629	1,463	8,093	10,875	18,581
2006	2,633	2,374	1,470	99	794	295	375	6,575	1,465	8,040	10,899	18,564

	Crop protection chemicals	Crop seed & biotech.	Farm machinery	Fertilizer	Food animal health1	Animal breeding & genetics2	Animal nutrition	Total crop inputs	Total animal inputs	Total agricultural inputs	Food manufacturing	Total food & agricultural inputs3
2007	2,676	2,540	1,618	101	793	297	378	6,934	1,468	8,402	11,152	19,176
2008	2,864	2,941	1,905	91	913	301	381	7,802	1,595	9,396	n.a.	n.a.
2009	2,814	3,149	2,176	94	876	308	382	8,232	1,566	9,799	n.a.	n.a.
2010	2,908	3,477	2,234	93	878	316	383	8,711	1,577	10,288	n.a.	n.a.

n.a. = not available. Current expenditures adjusted for inflation by the U.S. implicit Gross Domestic Product price deflator

[1]Animal health R&D is for food animals only, excluding R&D for companion and equine animal health.

[2]Estimates of private animal genetics research spending are only available for 1996 and 2006. We extrapolate for other years assuming 5.24 percent annual growth.

[3]Includes Organisation for Economic Development and Co-operation food industry R&D and total agricultural input R&D (animal nutrition is a subsector of the food industry and is not double counted in the total).

Source: USDA, Economic Research Service. See chapters for sources and estimation methods for specific industries.

Table 1.2. Public and private spending on food and agricultural research and development (R&D) worldwide in 2000

	Food R&D	Agriculture R&D	Food & ag R&D	Food & ag R&D[1]
Global total	Billion U.S. dollars			Billion PPP$
Public	n.a.	n.a.	16.3	20.8
Private	6.2	6.9	13.1	13.2
Total			29.3	33.9
Private share of total (%)			45.0	39.0
High-income countries				
Public	1.9(est)	7.4(est)	12.3	11.8
Private	5.8	6.3	12.2	11.8
Total	7.7(est)	13.7(est)	24.5	23.6
Private share of total (%)	76.0	46.0	50.0	50.0

n.a. = not available.

est. = estimate only. The allocation of public R&D into food-related and agriculture-related R&D in high-income countries is based on U.S. public R&D allocation shares and assumes these are roughly similar among all high-income countries. U.S. public R&D allocation is from the USDA's Inventory of Agricultural Research (USDA, 2000), which reports that in 2000, about 60 percent of total public agricultural R&D went to production agriculture, 15 percent went to food and nutrition, and the rest went to environmental and other topics. The total for public "food & ag R&D" includes all categories of research at public agricultural research institutions while the food and agricuilture sectors only include research directly related to that sector.

[1]The last column estimates international public R&D using purchasing-power-parity (PPP) exchange rates rather than the market exchange rates from which the U.S.$ estimates are derived. PPP exchange rates are based the relative price of a common basket of consumer goods. Using PPP exchange rates raises dollar estimates of R&D spending in developing countries significantly but affects spending estimates for high-income country only marginally. PPP exchange rates are from the World Bank.

Source: USDA, Economic Research Service. Estimates of public food and agricultural research are from Beintema and Stads (2008). Estimates of private food and agricultural R&D are from this study. Private R&D on animal nutrition is included in agriculture excluded from the food sector. See chapters for sources and estimation methods for specific industries.

Alston et al. (2010) also estimate that about 60 percent of U.S. public agricultural research was allocated to research relevant to farm productivity but do not provide a breakdown for the other 40 percent. If these figures are representative of public agricultural R&D in high-income countries, it would imply that the private sector accounts for roughly 76 percent of total food-related research and 46 percent of research on production agriculture in these countries (table 1.2).

For the biofuel industry, we estimate total private R&D at $1.47 billion in 2009 (table 1.3). This total includes $340 million spent by agricultural seed and biotechnology companies to improve biofuel feedstocks.[3] Another $1.03 billion was spent by companies in the energy sector to improve the efficiency of biofuel process manufacturing as well as to develop new types of biofuel feedstocks, such as algae. Enzyme and equipment manufacturers supplying inputs to energy companies for biofuel processing accounted for the remaining $71 million. Not included in these estimates is R&D spending by the transportation industry to modify vehicle and equipment engines for biofuel use. Although our estimates cover only one year, it is clear from industry sources that most of these R&D investments have arisen since 2000.

The largest driver of private biofuel R&D is the expectation of rising demand for alternative energy sources. This demand is sparked by the rising cost of fossil fuels relative to that for biomass-derived fuels and public concerns about national energy security and greenhouse gas emissions from fossil fuels. While government subsidies and regulations have helped stimulate demand for biofuel, public-sector investments in biofuel R&D now appear to be considerably less than private-sector investments. Moreover, business spending on biofuel R&D appears to be almost entirely from private capital: Government subsidies for private-sector biofuel R&D in the United States, historically the country with the largest government biofuel R&D program, amounted to only $24.4 million in 2009 (see chapter 10).

Table 1.3. Global expenditures for biofuel research and development (R&D) in 2009

Sector and type of firm	R&D
	Million U.S. dollars
Private sector market segments	
Agricultural input sectors (agricultural seed-biotechnology companies, plantations, forest product companies, and cellulosic biomass firms	340
Energy sector (biofuel producers, biofuel equipment manufacturers, and oil companies)	1,030
Enzyme and equipment input suppliers for biofuel processors	71
Total private biofuel R&D	1,470
Total public bioenergy R&D in industrialized countries	627

Source: USDA, Economic Research Service. For private R&D, see table 10.2. Public-sector bioenergy R&D is from the International Energy Agency. The 2009 total includes a one-time increase of $224 million in the United States due to the American Recovery and Reconstruction Act (economic stimulus funding).

R&D Spending by Region and for Selected Countries

Our estimates of private agricultural input R&D expenditures in specific countries or regions are based on the R&D expenditures by companies incorporated in that country or region.[4] The estimates of food industry R&D are based on national surveys of manufacturing enterprises as reported to the OECD, so they should reflect in-country R&D by domestic and foreign firms. While information on R&D spending by the food manufacturing industry is not available for most developing countries, our estimates include data for several, including China, Turkey, South Africa, Chile, and Mexico.

Among all countries in 2006, the United States was the leader in private food and agricultural R&D, accounting for about one-third of the global total (table 1.4). U.S. companies were particularly dominant in the crop seed/biotechnology and animal breeding sectors, accounting for about half of global private R&D in each sector. This high level of investments partly reflects the large U.S. domestic market for agricultural inputs, a strong and complementary public agricultural R&D system, and a relatively favorable regulatory environment for the commercialization of genetically modified (GM) crops (Fuglie et al., 1996). European firms accounted for about half of total R&D by agricultural input industries and just over a third of total R&D by the food industry (with Germany, Switzerland, and the

Netherlands being the leading countries in this region). Japan led R&D in the Asia-Pacific region. Japan had the second highest amount of R&D spending in the food industry (after the United States). In the agricultural input industries, Japan was among the leading countries in investing in R&D in the agricultural chemical and farm machinery sectors.

Table 1.4. Private-sector expenditures for food and agriculture research and development (R&D) by region in 2006

Sector	North America		Europe-Middle East	Asia-Pacifi c		Latin America	Global total
	All	United States		All	Japan		
	Million U.S. dollars						
Crop protection chemicals	599	599	1,596	404	368	34	2,633
Crop seed	1,287	1,261	983	96	66	6	2,374
Fertilizers	28	19	33	35	1	3	99
Farm machinery	573	513	579	309	189	9	1,470
Animal health[1]	279	236	477	36	8	3	794
Animal nutrition	66	63	232	71	19	7	375
Animal breeding	147	132	144	5	0	0	295
Crops	2,486	2,392	3,191	844	623	52	6,575
Animals	491	432	852	111	28	10	1,465
All agriculture	2,978	2,824	4,043	955	651	62	8,040
Food industry[2]	3,400	3,267	3,692	3,735	2,808	73	10,899
Food & agriculture[3]	6,312	6,028	7,503	4,619	3,440	128	18,564

[1]Animal health R&D includes R&D for food animals only. Globally, we estimate that food animal health R&D made up about 60 percent of total animal health R&D in 2006, based on the percentage of animal health product sales for food animals (see chapter 6).

[2]Food industry R&D is mainly for Organisation for Economic Co-operation and Development countries only.

[3]Sum of food industry R&D and all agriculture R&D. Animal nutrition is a subsector of the food industry and is counted in both food industry R&D and agricultural R&D but not in the total.

Source: USDA, Economic Research Service. See chapters for specific sources and methods.

Table 1.5 presents historical data on R&D spending by U.S. food processing and agricultural input industries. These time series data are reasonably complete for the food manufacturing, agricultural chemical, farm machinery, and animal health sectors. Estimates of R&D spending by the crop seed-biotechnology sector are available for 1993 onwards and for occasional earlier years but enough to establish a trend. R&D data are limited for fertilizer, animal nutrition, and animal genetics, but relatively little R&D is conducted by private companies in these sectors. The available data are sufficient to clearly show substantial growth in private food and agricultural R&D in the United States over the past three decades. Between 1979 and 2006—2 years with R&D estimates for all sectors—R&D spending by the private sector in the food and agricultural sectors increased more than fourfold (and more than doubled, from \$2.86 billion to \$6.03 billion, when viewed in constant 2006 U.S. dollars), although this growth is less than that for U.S. industry generally. By comparison, total R&D funded and performed by all U.S. private industries increased nearly ninefold, from \$25.6 billion to \$223.4 billion (nominal dollars), over the same period (NSF, 2010).

Table 1.5. Private food and agricultural research and development (R&D) spending in the United States

Year	Crop protection chemicals	Crop seed & biotech	Fertilizers	Farm machinery	Animal health (all animals)	Animal health (food animals only)	Animal nutrition	Animal genetics	Food industry
				Million nominal U.S. dollars					
1960	27	4		75	6				104
1961	38			65	11				
1962	42			70	13				121
1963	45			76	15				130
1964	48			79	20				144
1965	64	6		96	23				
1966	77			100	28				164
1967	92			102	35				181
1968	99			96	36				184
1969	104			99	34				
1970	126	11		89	45				222
1971	130			90	48				238
1972	108			93	53				258
1973	114			120	62				268
1974	137			131	74				297
1975	176	24	3	138	79		28		335
1976	205			168	87				355
1977	236			221	84				415
1978			3		86		30	44	472
1979	292	43	3	295	96		33	55	528
1980					111				620
1981	487			278	125				636
1982		115			129				777
1983	587			290	147				824
1984			22		154		42		1,081
1985	432			368	159				1,136
1986					179				1,280
1987	398			483	191				1,206
1988					221				1,229
1989	561	272		281	243				1,275
1990					245				1,414
1991	614			413	276				1,277
1992					331				1,386
1993	686	409		276	315	176			1,345
1994	707	425		302	244	134			1,476
1995	751	507		361	337	182			1,566
1996	834	636		471	342	181	49	118	1,564
1997	897	791		507	353	183			1,908
1998	847	963		520	369	188			1,949
1999	756	991		371	374	187			1,563
2000	703	1,045		420	358	175			1,562
2001	531	985		395	349	168			1,971
2002	534	1,010		372	342	161			2,204
2003	558	1,012		403	417	192			2,160
2004	606	1,078		453	478	215			2,809
2005	612	1,095		504	524	230			3,255
2006	599	1,261	19	513	549	236	63	132	3,267

Table 1.5. (Continued)

Year	Crop protection chemicals	Crop seed & biotech	Fertilizers	Farm machinery	Animal health (all animals)	Animal health (food animals only)	Animal nutrition	Animal genetics	Food industry
				Million nominal U.S. dollars					
2007	614	1,393		628	641	269	74		2,939
2008	683	1,707		813	830	340	92		n.a.
2009	740	1,897		1,057	783	313	71		n.a.
2010	793	2,176		1,120	772	309			n.a.

Sources: USDA, Economic Research Service. For 1993-2007 continuous time series, see individual chapters for detailed sources and estimation procedures. For pre-1993 data: crop seed research: 1960-1979 (Perrin et al., 1983); 1982 (Kalton and Richardson, 1983); 1989 (Kalton et al., 1989). Animal genetics research: 1978-79 (Malmstead, as reported in Ruttan, 1982); 1996 (Narrod and Fuglie, 2001). Agricultural chemicals, farm machinery, and food industry (NSF, various issues). Animal health (Pharmaceutical Research and Manufacturers of America, annual reports).
Fertilizer and/or animal nutrition: 1975 (Wilcke and Williamson, 1977); 1978-79 (Malmstead, as reported in Ruttan, 1982); 1984 (Crosby, 1987), 1996 (Fuglie et al., 2000).

Private spending on food and agricultural R&D in the United States has exceeded public-sector agricultural research expenditures most years since the late 1970s (fig. 1.1). Federal and State governments invested on average $4.40 billion annually (constant 2006 dollars) in agricultural research between 1980 and 2007, while the private sector spent an average of $4.95 billion per year (constant 2006 dollars) over the same period. But as previously discussed, each sector focuses its research resources differently. The private sector accounts for about 80 percent of total food-related research and about 47 percent of total research related to production agriculture. Within these areas, public research is more oriented toward basic or fundamental science and scientific training, as well as topics like food safety, genetic resource conservation, and farming practices to conserve natural resources, research that has high social value but for which private incentives are relatively weak.

Market Size and Private Food and Agricultural R&D

One key determinant of private investment in R&D is the size of the market for products or processes developed from the R&D. Sales of new products or cost savings from manufacturing (process improvements) are necessary for firms to earn a return from their R&D.

Moreover, firms must be able to price new products above their cost of manufacture, at least for some period of time, to help recoup the sunk costs of R&D, regulatory approval, and market development.

Appropriability is the ability of firms to exercise some market power in the marketing and pricing of new products derived from their R&D investments.

Securing patents and other forms of intellectual property rights enables firms to exercise appropriability over the economic benefits provided by the application of new technology.

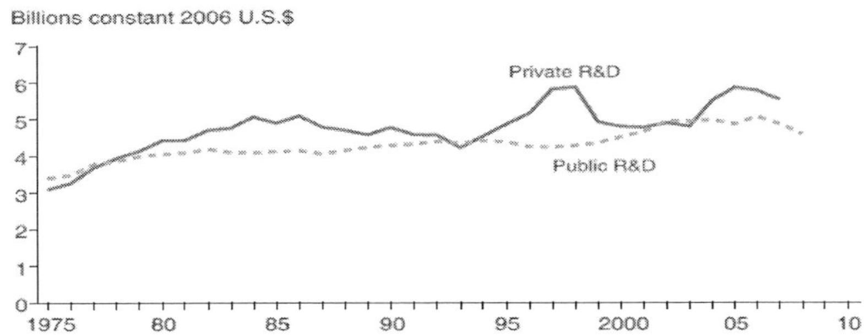

Source: U.S. public agricultural research and development (R&D) spending is from USDA, Economic Research Service. U.S. private R&D spending is derived from the data in table 1.5, with interpolations for missing data. Nominal research expenditures are adjusted for inflation by the agricultural R&D price index developed by ERS. This price index takes into account changes in the cost of research inputs (scientist salaries, scientific equipment, etc.).

Figure 1.1. Trends in public and private food and agricultural research spending in the United States.

Global Demand for Agricultural Inputs

Information on the size of global markets for agricultural inputs is not readily available. Thus, we have assembled data from a variety of sources or made estimates of the wholesale value of market sales for agricultural inputs by product type. In 2006, total company sales of these inputs were $355 billion (table 1.6). Fertilizers and animal feed (not including medicated feeds, which we include in the animal health sector) are the largest markets in terms of sales and consist of mostly bulk inputs that do not involve much R&D. These products accounted for about 60 percent of total agricultural input sales. Another 21 percent was for farm machinery and equipment. Crop protection chemicals and crop seed together accounted for about 15 percent of inputs purchased by farmers, while animal health and breeding materials accounted for the remaining 4 percent. Measures of the size of the various input markets vary somewhat depending on the source. Estimates of private-sector sales of crop seed and animal breeding materials vary the most. Historically, farmers have met a portion of their demand for crop seed and animal breeding stock through self-supply or by obtaining these inputs through informal markets or from neighboring farms. Over time, specialized breeding firms have increasingly helped meet this demand. By 2006, private seed companies appeared to be supplying about two-thirds of the crop seed used globally. The private-sector share of animal breeding stock is not known with much precision but appears to be very high for poultry; high and rising for swine and dairy cattle; and relatively low for beef cattle, small ruminants, and aquaculture (with the exception of some species, such as salmon). A comparison of private-sector sales of farm machinery, crop protection chemicals, crop seed, and food animal health products worldwide since 1994 shows that only the markets for farm machinery and crop seed have grown significantly in inflation-adjusted dollars (fig. 1.2). Global sales of crop protection chemicals recovered somewhat from their low in 2002 but only to mid-1990s levels (to some extent, the increasing use of GM crops with pesticidal properties may be substituting for chemicals in crop protection). Most of the growth in sales of animal health products was attributed to markets for nonfood animal species, such as companion and equine animals. The figure does not show market trends for the animal feed and fertilizer markets. Although these are the largest agricultural input markets (in terms of sales), they are mostly characterized by bulk, homogeneous products and little private R&D. Data are unavailable for trends in commercial sales of animal genetics products.

Table 1.6. Global market for agricultural inputs supplied by the private sector in 2006

Industry	Segment	Privatesector sales
		Million US$
Crop protection chemicals	Total for agricultural uses	31,962
	Herbicides	15,246
	Insecticides	7,895
	Fungicides	7,671
	Other	1,151
Crop fertilizers	Total (168 million tons)	74,692
	N fertilizer (99 million tons)	48,076
	P_2 O_5 fertilizer (39 million tons)	17,875
	K_2O fertilizer (30 million tons)	8,741
Crop seed	Total proprietary seed sales	19,600
	Conventional seed (proprietary)	11,800
	Genetically modifi ed seed (proprietary)	7,800
	Public seed sales and farmer-saved seed (not included in total)	9,400
Farm machinery	Total	73,579
	Farm tractors	21,321
	Harvesting machinery	16,455
	Planting and fertilizing machinery	35,802
Animal health	Total for food animals	9,455
	Total (food, companion and equine animals)	16,065
	Pharmaceuticals	10,410
	Biologicals (vaccines)	3,660
	Medicated feed additives	1,995
Animal nutrition	Total	141,833
	Compound feed (656 million tons)	137,429
	Nutritional feed additives	4,404
	Medicated feed additives	(see animal health)
Animal breeding	Total	4,062
	Poultry	1,742
	Pigs	1,303
	Cattle	931
	Aquaculture	87
All private-sector sales of farm inputs		355,182

Sources: USDA, Economic Research Service. Agricultural chemicals from AGROW Reports (2007); crop seed sales from Context Network (2007); fertilizer sales derived from quantities of nutrients reported in Food and Agricultural Organization multiplied trade prices (dollars per metric ton of nutrient) from Haver Analytics; animal health products from Vetnosis as reported in International Federation for Animal Health (2007); animal feed sales derived the quantities reported in Best (2008) multiplied by International Monetary Fund corn and soy meal prices adjusted for processing costs; animal breeding are authors' estimates (see chapter 7); farm machinery from Freedonia (2006).

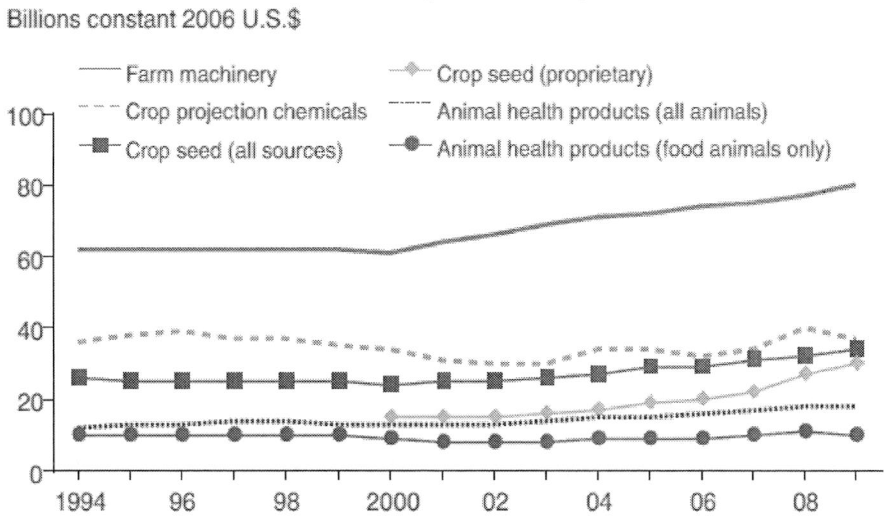

Source: USDA, Economic Research Service. See sources listed in notes to table 1.6.

Figure 1.2. Global market sales of selected agricultural inputs.

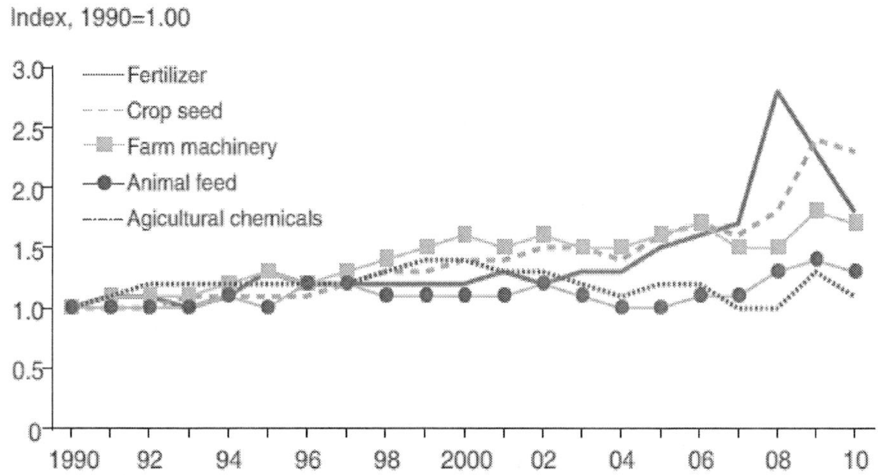

Source: USDA, Economic Research Service. Indexes of prices paid and received by farmers from USDA (various issues).

Figure 1.3. U.S. agricultural input prices relative to prices received by farmers.

Price Trends for Some Agricultural Inputs

Markets for agricultural inputs can expand through either larger volumes of sales or through higher unit prices. Upward trends in unit prices may reflect rising quality of inputs, such as new technologies embodied in the inputs due to past investments in R&D. Higher input prices may also stem from increases in manufacturing costs due to rising labor, capital, or material costs. Based on a comparison of five categories of agricultural input prices received by farmers in the United States,[5] the largest change during 1994-2010 was in crop seed prices, which more than doubled relative to the price received for agricultural commodities sold by farmers (fig. 1.3). This increase was due, at least in part, to the increase in value-added characteristics developed by private seed and biotechnology companies through R&D programs. Le Buanec (2008) estimates that between 32 and 74 percent of the price of seed for corn, soybeans, cotton, and sugar beets in the United States and the European Union (EU) reflects technology fees or the cost of seed treatments. The sharp rise in the price of fertilizer in 2008-09 was driven by a significant increase in the cost of energy and materials used to manufacture fertilizer (especially natural gas, sulfur, and phosphate rock), as well as an increase in transportation costs and the falling value of the U.S. dollar (Huang, 2009). For agricultural chemicals, prices rose relative to commodity prices during 1994-99 but have since fallen. The recent decline partly reflects the rise in crop commodity prices after 2005 as well as an increasing market share for off-patent (generic) crop protection chemicals.

Market Structure and R&D in Agricultural Input Industries

The growth rates in the global market size for agricultural inputs is generally consistent with the trends in private spending on agricultural input R&D (see table 1.1), with the important exception of crop seed-biotechnology, where R&D grew more rapidly than sales value. We generally expect research investments to be correlated with industry sales (i.e., that *research intensity*, or the R&D-to-sales ratio, remains stable over time) unless other factors are changing incentives for private R&D. Other factors include (1) expectations that future demand growth will accelerate, (2) advances in scientific knowledge that have created new technological opportunities for commercialization, and (3) stronger IPR or changes in market structure that have made it easier for private R&D investors to appropriate economic benefits of new technology. Greater industry concentration, like stronger IPR, can increase appropriability if it strengthens the market power of large firms. Market power enables firms to charge more for new or existing products and recoup their sunk investments in R&D and market development. These factors may not be acting separately but may be working concurrently to change incentives for private R&D. For example, scientific advances in molecular biology have created new technological opportunities in agricultural biotechnology and changes in IPR have increased appropriability over biological innovations (Fuglie et al., 1996). In such an environment, firms may consolidate to acquire complementary technology and marketing assets, capture economies of scale in R&D, and strengthen their market power. Indeed, across a number of agricultural input industries, mergers, acquisitions, and consolidation among firms are affecting industry concentration and structure.

Changes in Industry Concentration and R&D Intensity over Time

In each of the five agricultural input industries with significant R&D, the degree of concentration in the global market rose significantly over 1994- 2009, although a lack of data prevented us from quantifying this change for the animal breeding sector (table 1.7). We measure concentration using the Herfindahl index and by four-firm and eight-firm concentration ratios.[6] By the end of the present decade, the largest four firms accounted for at least 50 percent of global market sales in each of these five agricultural input sectors. By 2006/07, market concentration was particularly high in the animal breeding sector, where the four-firm concentration ratio reached 56 percent. Growth in market concentration over time was most rapid in the global seed industry, where the market share of the four largest firms more than doubled from 21 to 54 percent between 1994 and 2009.

Table 1.7 also shows the trend in R&D intensity (i.e., R&D spending as a percentage of sales) for each agricultural input industry. With the exception of the crop seed-biotechnology sector, R&D intensity for each sector remained fairly constant over 1994-2009, although it varied significantly across sectors. R&D intensities averaged 8.6 percent for the animal health industry, 6.7 percent for the agricultural chemical industry, and 2.3 percent for the farm machinery industry.

For the crop seed industry, R&D intensity increased from 11.0 percent in 1994 to 15.0 percent in 2000 and then fell back to 10.5 percent by 2009. For the animal breeding sector, we have an estimate of R&D intensity for 2006/07 only: an average of 7.3 percent across species.

Table 1.7. Market concentration and research and development (R&D) intensity in global agricultural input industries

Year	Herfindahl index	4-firm concentration ratio	8-firm concentration ratio	Industry R&D intensity
Crop protection chemicals		*Share of market (%)*		*R&D/sales (%)*
1994	398	28.5	50.1	7.0
2000	645	41.0	62.6	6.8
2009	937	53.0	74.8	6.4
Crop seed and traits				
1994	171	21.1	29.0	11.0
2000	349	32.5	43.1	15.0
2009	991	53.9	63.4	10.5
Animal health				
1994	510	32.4	57.4	8.6
2000	657	41.8	67.4	8.5
2009	827	50.6	72.0	8.6
Farm machinery				
1994	264	28.1	40.9	1.9
2000	353	32.8	44.7	2.3
2009	791	50.1	61.4	2.7
Animal genetics				
1994	n.a.	n.a.	n.a.	n.a.
2000	n.a.	n.a.	n.a.	n.a.
2006/07	1,025	55.9	72.8	7.3

n.a. = not available.

Source: USDA, Economic Research Service estimates based on firm-level sales and R&D expenditure data collected for this study. See specific chapters for details.

Greater concentration was not associated with a permanent rise in R&D intensity in these input industries. In the crop seed industry, there was a temporary increase in research intensity in the late 1990s and early 2000s as the industry sought to commercialize a number of genentically modified crop varieties.

But by the late 2000s, research intensity in the crop seed industry was back to its mid-1990s level. In fact, the underlying causes of growing concentration in these sectors appear to be quite specific to each sector and may not have affected private incentives to invest in R&D (table 1.8).

Table 1.8. Factors driving changes in market structure in global agricultural input industries

Sector	Factors driving consolidation and concentration	Change in real R&D spending between 1994 and 2010[1]
		Percent
Crop seed & biotechnology	Acquisition of complementary technology and marketing assets, economics of scale in crop biotechnology R&D	138
Farm machinery	Financial losses of major manufacturers during farm sector business cycles (which strongly infl uence demand for large capital purchases)	88
Animal breeding & genetics	Vertical integration of poultry and livestock industries; economics of scale in animal biotechnology R&D	25
Animal health (food animals only)	Forces driving consolidation in the pharmaceutical industry: loss of profi t streams and idled capacity when major drugs go off-patent	2
Crop protection chemicals	Stricter environmental and safety regulations; maturing markets; rise of generic products	-2

[1] We have data on research and development (R&D) spending by the animal breeding and genetics industry for 1996 and 2006/07 only. The estimate of 25 percent growth between 1994 and 2010 is derived by applying the 1996-2006 average annual growth rate to these years. Changes in real R&D spending calculated from the data in table 1.1.

Source: USDA, Economic Research Service. See chapters for discussion of specific industries.

In the crop seed and animal breeding sectors, the emergence of biotechnology was a major driver of consolidation. Firms sought to acquire relevant technological capacities and serve larger markets to spread the large fixed costs associated with meeting regulatory approval costs for new biotechnology innovations. In the poultry and livestock sectors, vertical integration enabled some large firms to acquire capacity in animal breeding as part of their integrated system. In the farm machinery industry, many of the major mergers and acquisitions can be traced to large financial losses sustained by some leading firms during periods in which the farm sector was in prolonged recession, which substantially reduced demand for farm machinery as farmers delayed major capital purchase. Firms experiencing large financial losses are often vulnerable to acquisition. The crop protection sector has been heavily affected by changes in regulations governing the health, safety, and environmental impacts of new and existing pesticide formulations. The consolidation in the animal health sector appears to be largely a byproduct of mergers and acquisitions in the pharmaceutical industry (as most of the leading animal health companies are subsidiaries of large pharmaceutical companies).

R&D Spending by Firm Size

Large firms usually account for most of the R&D spending in an industry. They may have, on average, higher R&D-to-sales ratios than smaller firms. If R&D-oriented large firms acquire small firms that do not make considerable investments in R&D, such consolidation could lead to greater R&D by the industry as a whole. On the other hand, mergers between R&D-oriented firms could reduce overall R&D spending as duplication and redundancies in their merged R&D programs are eliminated. Merger activity may also be led by firms that specialize in off-patent generic products. A growing market share by these firms may lead to lower R&D in the industry as a whole. But the results reported earlier suggest that with the exception of the crop seed-biotechnology industry, market consolidation has generally not been correlated with changes in overall R&D by the sector.

An examination of average R&D intensities, global R&D shares, and global market shares for different classes of firms in four agricultural input sectors reveals trends between R&D and firm size (table 1.9). The general pattern is for four to eight of the largest firms to have the highest R&D-to-sales ratio and account for most R&D by the sector. For crop protection chemicals, five large, research-oriented ("discovery") firms accounted for 74 percent of total R&D and 57 percent of total market sales for this sector. Another group of 17 midsized firms also invested in the discovery of new proprietary products and accounted for most of the rest of the R&D related to agricultural chemicals.

Table 1.9. Company size and research and development (R&D) spending in agricultural input industries in 2006

Sector	Companies	Average R&D intensity	Global R&D share	Global market share
Crop protection chemicals	—— *Number* ——		—— *Percent* ——	
Large discovery companies (>$2 billion sales)	5	9.0	74.1	57.4
Second-tier discovery companies (<$2 billion sales)	17	7.3	19.6	18.7
Other manufacturers	23	2.3	7.7	23.9 est.
Crop seed and biotechnology				
Large seed companies (> $600 million sales) + BASF	8	15.8	75.6	48.8
Midsize seed companies ($50-600 million sales)	29	7.3	13.7	19.2
Other seed companies	n.a.	2.0	3.1	16.0 est.
Agricultural biotechnology companies	58	42.1	7.6	1.8
Animal health Large animal health discovery companies (>$800 million in sales)	8	10.0	66.7	79.6
Midsize animal health companies ($250 million-$800 million sales)	5	7.6	11.8	10.6
Other manufacturers	n.a.	3.8	21.5	9.8 est.
Farm machinery				
Leading multiline farm machinery companies (>$5 billion sales)	4	3.0	57.4	38.7
Second-tier farm machinery manufacturers	30	2.4	27.6	22.9
Other manufacturers	n.a.	2.4	0.6	0.5 est.

est. = authors' estimate. n.a. = not available.

Source: USDA, Economic Research Service. See chapters for specific sources and methods.

The average R&D intensity for the smaller sized firms was slightly below that of the largest. Generic producers (firms not investing in new product discovery) conducted a small amount of R&D related to product manufacture and registration. In the crop seed-biotechnology sector, the largest eight seed sellers plus BASF (a firm investing significantly in agricultural biotechnology R&D but with few direct seed sales) accounted for 76 percent of private-sector seed research and had an average R&D intensity more than double that of midsized seed firms. However, small agricultural biotechnology firms had by far the largest research intensity in this sector, at about 42 percent. These operations tend to be startup organizations seeking to commercialize new research discoveries. If they are successful, they are likely to partner with large seed-biotechnology firms or be acquired by one of them. They play an important role in bringing high-potential but high-risk technologies into the marketplace. In the animal health and farm machinery sectors, the leading firms also had the highest average R&D intensities. (A number of biotechnology firms are conducting research on animal health, but few specialize in the agricultural sector and none are included in table 1.9.)

Globalization of Private Agricultural R&D

All of the leading firms and many of the second-tier firms in food manufacturing and agricultural input industries are multinational, offering product sales spread across several continents. In fact, global trade in agricultural inputs has grown rapidly over the past two decades (table 1.10). Between 1990 and 2007, international trade in animal breeding material grew by 260 percent and trade in farm machinery grew by 190 percent (in constant 2006 U.S. dollars). Trade in crop protection chemicals and crop seed also grew over the period (trade statistics for animal health products are not available).

Table 1.10. Global trade in agricultural inputs

Input type	Value of global exports		
	1990	2000	2007
	Billion constant 2006 U.S.$		
Farm machinery	24.1	33.3	69.6
Crop protection chemicals	10.6	13.0	18.2
Crop seed	4.1	4.3	6.0
Animal breeding material	0.3	0.5	1.2

Sources: USDA, Economic Research Service. Farm machinery and pesticide export values from Food and Agriculture Organization; Crop seed export value from the Le Buanec (2007) and International Seed Federation; trade in animal breeding material includes value of exports of day-old poultry chicks, swine and bovine live breeding animals, and bovine semen (UN ComTrade). Export values adjusted for inflation by the U.S. Gross Domestic Product implicit price deflator (*Economic Report of the President, 2009*).

Since the performance of agricultural technologies tends to be site specific (due to variations in weather, soil type, and other environmental conditions), many of the leading agricultural input firms have located R&D facilities around the world. This global R&D presence not only allows firms to develop and adapt new technologies to regional conditions and meet local regulatory requirements, but it also may enable them to achieve cost economies in some R&D activities (e.g., by conducting certain kinds of research in countries where highly trained personnel or specialized R&D services can be hired more cheaply). While we do not have direct information on R&D investment in foreign countries by these firms, we have assembled information on the global R&D presence for several of the leading agricultural input firms (see table 1.11).

Table 1.11. Agricultural research and development (R&D) spending by major multinational corporations and public institutions in 2007

Company	Country of incorporation	Sector of R&D activity	Agricultural R&D spending (estimate only)	Principal agricultural R&D locations
Million U.S.$				
Bayer[1]	Germany	Ag. chemical, crop seed, animal health	978	Germany, France, Belgium, Netherlands, U.S., Japan
Syngenta[2]	Switzerland	Ag. chemical, crop seed	830	Switzerland, UK, U.S., China, Australia
Monsanto[3]	U.S.	Ag. chemical, crop seed	770	U.S., France, Brazil, Argentina, India, Australia
BASF[4]	Germany	Ag. chemical, crop seed, animal nutrition	655	Germany, U.S., India
Dupont[5]	U.S.	Ag. chemical, crop seed, food ingredients	633	U.S., France, Japan, India
Dow[6]	U.S.	Ag. chemical, crop seed	294-380	U.S., Japan, Argentina, Puerto Rico
Limagrain[7]	France	Crop seed	171	EU, U.S., Brazil, Chile, China, Japan, Israel, Morocco
KWS[8]	Germany	Crop seed	104	EU, U.S., Argentina, China, Turkey, Russia
John Deere[9]	U.S.	Farm machinery	461	U.S., India, Israel
CNH[9]	Netherlands	Farm machinery	272	U.S., EU, Brazil, Turkey, India, China
CLAAS[10]	Germany	Farm machinery	150	Germany
Pfi zer[11]	U.S.	Animal health	317	U.S., UK, Japan
Merial[11]	U.S. & UK	Animal health	250	U.S., France, 9 global locations
Schering-Plough[11]	U.S.	Animal health	113	U.S., 14 global locations
Fort Dodge (Wyeth)11	U.S.	Animal health	115	U.S., EU
DSM[12]	Netherlands	Animal nutrition	114	Netherlands
Genus[13]	UK	Animal genetics	33	U.S., UK
Public – USDA/ARS[14]	U.S.	Crop science	456	U.S.
Public – USDA/ARS[14]	U.S.	Animal science	171	U.S.
Public – CGIAR[14]	Global	Agricultural biodiversity and genetic improvement	178	9 centers with crop breeding programs, all in developing countries

[1]Bayer reports spending 506 million euros on crop protection R&D and 131 million euros on environment science/bioscience in 2007 (bioscience is mostly seed and crop biotechnology research while environmental science includes nonagricultural applications of crop protection chemicals and related products). Since 2006, Bayer no longer reports animal health R&D separately from its Consumer Health business segment, but it did report animal health product sales of 956 million euros in 2007. We estimate Bayer spent 8 percent of animal health sales on R&D, or 76 million euros. These figures are from Bayer (2008). [2]Syngenta reports spending $496 million on crop protection R&D, $283 million on crop seed R&D, and $51 million on new business development (mostly crop biotechnology) R&D in 2007 (Syngenta, 2008). 3 Monsanto

reports spending $770 million on agricultural R&D in 2007, mostly for its seeds and genomics division, with the remainder to support its crop protection products (Monsanto, 2009). [4]BASF (2007) reports that the company spent 328 million euros on crop protection R&D in 2007 and 400 million euros on plant sciences R&D over 2006-08 (the latter is included as part of its corporate"Verbund" research for future business development). We assume it spent about one-third of this 3-year total, or 135 million euros, for plant sciences R&D in 2007. In addition, BASF develops animal nutrition specialty products (vitamins, enzymes, and minerals). It does not report animal nutrition sales or R&D separately but includes this in its fine chemicals business segment, although for 2009 it reported that animal nutrition sales made up 16 percent of product sales from this segment (BASF, 2010). We assume animal nutrition products accounted for 16 percent of sales of fi ne chemicals in 2007 (485 million euros) and that BASF invested 3 percent of this, or 15 million euros, in animal nutrition R&D in 2007. [5]Dupont (2008) reports that its agriculture and nutrition division spent $633 million on R&D in 2007. Net sales from this business segment included crop seeds (49 percent), crop protection chemicals (34 percent), and food ingredients (17 percent). [6]Dow does not report R&D spending by business segment but is known to invest signifi cantly in both crop protection and crop seed and biotechnology R&D. In 2007, Dow's total R&D spending was $1,305 million (Dow Chemical Co., 2009). We derive a lower bound estimate of Dow's agricultural R&D spending by multiplying total R&D by the share of agricultural science patents in Dow's total U.S. patent holdings, which were 508 out of 2,266 patents as of December 31, 2008, according to Dow Chemical Co. (2009). Our upper bound estimate is derived assuming Dow invested 10 percent of its crop protection sales and 33 percent of its seed sales in R&D. While this research intensity for seed is high, it reflects Dow's stated intention to expand its market presence in the global seed industry. [7]Limagrain spent 102 million euros in crop seed research in 2006/07 (Limagrain, 2007). [8]KWS spent 75 million euros in crop seed research in 2006/07 (KWS, 2008). [9]John Deere and CNH report total spending for research, development and engineering for agricultural, construction, and other equipment sales. We estimate their R&D spending for agricultural equipment by taking the proportion of agricultural sales in total equipment sales. For Deere, this implies 56 percent of its total R&D spending of $817 million was for agriculture in 2007 (Deere & Company, 2007) and for CNH, 66 percent of total R&D spending of $409 million was for agriculture in 2007 (CNH, 2008). [10]CLAAS reports spending 110 million euros on research, development, and engineering for agricultural equipment in 2007 (CLAAS, 2009). [11]These pharmaceutical companies do not report animal health R&D separately, although they do report animal health product sales. To estimate animal health R&D for these countries, we use estimates of R&D as percentage of animal health sales as reported in Animal Pharm Reports (2007). These are: 12 percent for Pfi zer, 10 percent for Merial and Fort Dodge, and 9 percent for Schering-Plough. See chapter 6 for recent merger activity in animal health. [12]DSM develops and markets both animal and human nutrition and health products. Its total R&D spending in 2007 was 136 million euros. We assume that 57 percent of this was for animal nutrition R&D, the same proportion of animal product sales out of total nutrition sales. (DSM, 2007). [13]Genus reports 17.7 million euros in R&D spending for livestock (cattle and pigs) research in 2007 (Genus, 2007). [14]For comparative purposes, we show agricultural R&D spending for two prominent public-sector institutions: USDA's Agricultural Research Service (USDA/ARS) and the research centers that are supported by the Consultative Group for International Agricultural Research (CGIAR). USDA/ARS expenditures for crop and animal sciences are from USDA (2007); CGIAR spending on biodiversity conservation and genetic improvement (which is mostly for food crops) is from CGIAR (2007).

We convert foreign currencies into U.S. dollars using the exchange rates reported in the Economic Report of the President (2009). Sources: USDA, Economic Research Service and others, as noted above.

Based on information from company websites, we indicate the sectors in which these firms made R&D investments in 2007 and the countries or regions of their principal agricultural R&D facilities. In addition to these principal research locations, the companies may have had field-testing stations and manufacturing facilities in several other countries. For comparative purposes, the last three rows of table 1.11 shows R&D spending by some of the largest public-sector agricultural research institutions. It is noteworthy that at least five firms made larger investments in crop improvement than the world's largest public-sector agricultural research agency, USDA's Agricultural Research Service (ARS), and several times the investment in crop genetic conservation and breeding than the network of centers that make up the Consultative Group for International Agricultural Research (CGIAR).[7] The three companies that made the largest investments in agricultural research in 2007 were the European firms Bayer and Syngenta and the U.S. firm Monsanto, each with over $700 million in R&D spending for crop and/or animal agriculture. By 2007, the agricultural R&D investment by these three firms together was $2.47 billion (and it rose further to over $3 billion by 2009[8]).

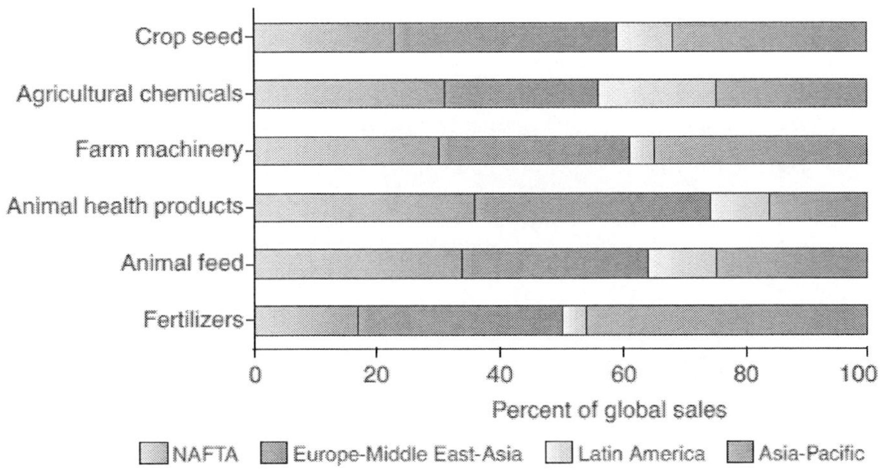

Note: Global distribution of sales of animal genetics is not available.
Source: USDA, Economic Research Service. See chapters for sources on specific input industries.

Figure 1.4. Global distribution of agricultural inputs sales in 2006.

Another indicator of the degree of globalization of agricultural input markets is the global distribution of agricultural input sales (see fig. 1.4). In 2006, member countries of the North American Free Trade Agreement (NAFTA—United States, Canada, and Mexico) accounted for about 23 percent of the global seed market and 30-36 percent of global sales of agricultural chemicals, farm machinery, animal feed, and animal health pharmaceuticals (including those for nonfood animals). The Europe-Middle East-Africa market (which is mostly Europe) had the largest aggregate seed sales in 2006, whereas Asia-Pacific countries used the most fertilizers and bought the most farm machinery. Together, the Asia-Pacific and Latin America examples are indicative of a rough estimate of the developing-country share of global agricultural input markets.[9] They account for 37-51 percent of global sales of crop seed and chemicals, farm machinery, fertilizers, and animal feed.

These indicators—trade in agricultural inputs, location of R&D facilities, and the wide distribution of agricultural input sales—demonstrate the multinational nature of private-sector investments in agricultural R&D and the role of these companies in developing and transferring agricultural technology around the world. One implication of the globalization of private-sector food and agricultural research is that the rate of international technology transfer may accelerate, eventually serving to reduce productivity differences across nations and regions. Moreover, the location of principal R&D centers may be less important than the location of markets and flow of trade in the agricultural inputs that embody the technology developed through this R&D.

References

AGROW Reports. 2007. GM Crops, Informa UK Ltd., London, July.

Alston, J., M. Anderson, J. James, and P. Pardey. 2010. *Persistence Pays: U.S. Agricultural Productivity Growth and the Benefits From Public R&D Spending*, New York: Springer.

Animal Pharm Reports. 2007. *Animal Pharms' Top Tier Companies: 2007 Edition*, Informa UK Ltd., London, September.

BASF. 2007. *BASF Annual Report 2007*, BASF SE, D-67056 Ludwigshafen, Germany.

BASF. 2010. *BASF Factbook 2010,* BASF SE, D-67056 Ludwigshafen, Germany, June.

Bayer. 2008. *Bayer Annual Report 2007*, Bayer AG, 51368 Leverkusen, Germany, February 28.

Bayer. 2010. *Bayer Annual Report 2009*, Bayer AG, 51368 Leverkusen, Germany, February 26.

Beintema, Nienke M., and Gert-Jan Stads. 2008. "Measuring Agricultural Research Investments: A Revised Global Picture," *ASTI Background Note, Agricultural Science and Technology Indicators*, International Food Policy Research Institute.

Best, Peter. 2009. "World Feed Panorama," *Feed International*, January, pp. 12-15.

CLAAS. 2009. CLAAS 2009 Annual Report, CLAAS KGaA mbH, D-33428 Harsewinkel, Germany.

CNH. 2008. Form 20-F for 2007, CNH Global NV, Amsterdam, Netherlands, March 5.

Cohen, Wesley M., Richard C. Levin, and David C. Mowery. 1987. "Firm Size and R&D Intensity: A Re-Examination," *Journal of Industrial Economics* 35, 4 (June): 543-565.

Consultative Group for International Agricultural Research (CGIAR). 2007. *Financial Report 2007*, CGIAR Secretariat, World Bank.

Context Network. 2007. "Sizing Up the Seed Market," *Seed World*, pp. 10-12.

Crosby, Edwin A. 1987. "Private Sector Agricultural Research in the United States," in *Policy for Agricultural Research*, V.W. Ruttan and C.E. Pray (eds.), Boulder, CO: Westview Press, pp. 395-409.

Deere & Company. 2007. *Annual Report 2007*, Moline, IL.

Dow Chemical Co. 2009. Form 10-K, 2020 Dow Center, Midland, MI, February 20.

DSM. 2007. DSM Annual Report 2007, Royal DSM NV, 6401 JH Heerlen, Netherlands.

Dupont. 2008. *Dupont 2007 Databook*, Wilmington, DE.

Echeverria, Ruben G., Eduardo J. Trigo, and Derek Byerlee. 1996. *Institutional Change and Effective Financing of Agricultural Research in Latin America*, World Bank Technical Paper No. 330, World Bank.

Economic Report of the President. 2009. U.S. Government Printing Office.

Federal Reserve Bank of St. Louis. Producer Price Indexes, FRED Economic Data, available at: http://research.stlouisfed.org/fred2/categories/31 (accessed September, 2009).

Food and Agricultural Organization of the United Nations (FAO). FAOSTAT database, Rome, Italy, available at: http://faostat.fao.org/ (accessed October, 2010).

Freedonia. 2006. *World Agricultural Equipment to 2010*, Report Number 2089, Freedonia Group.

Frey, Kenneth J. 1996. *National Plant Breeding Study-I: Human and Financial Resources Devoted to Plant Breeding Research and Development in the United States in 1994*, Special Report 98, Iowa Agriculture and Home Economics Experiment Station, Iowa State University.

Fuglie, Keith O., Nicole Ballenger, Kelly Day, Cassandra Klotz, Michael Ollinger, John Reilly, Utpal Vasavada, and Jet Yee. 1996. *Agricultural Research and Development: Public and Private Investments Under Alternative Markets and Institutions*, Agricultural Economic Report 735, U.S. Department of Agriculture, Economic Research Service, available at: www.ers.usda.gov/publications/aer735/

Fuglie, Keith O., and D.E. Schimmelpfennig (eds.). 2000. *Public-Private Collaboration in Agricultural Research: New Institutional Arrangements and Economic Implications*, Ames, IA: Iowa State University Press.

Genus. 2007. *Annual Report and Financial Statements 2007*, Genus Plc, Basingstoke, Hampshire, United Kingdom.

Haver Analytics. Commodity Prices, International Financial Statistics. DLX Databases, Haver Analytics, Inc., available through www.haver.com/ (subscription required).

Huang, Wen-yuan. 2009. *Factors Contributing to the Recent Increase in U.S. Fertilizer Prices, 2002-08*, Agricultural Resources Situation and Outlook No. AR-33, U.S. Department of Agriculture, Economic Research Service, available at: www.ers.usda.gov/publications/ar33/

Huffman, Wallace E., and Robert E. Evenson. 2006. *Science for Agriculture: A Long-Term Perspective, 2nd edition.* Ames, IA: Iowa State University Press.

International Energy Administration. R&D Statistics, IEA/OECD, Paris, France, available at: http://data.iea.org/ (accessed November 30, 2010).

International Federation for Animal Health. 2007. *IFAH Annual Report 2008*. Brussels, Belgium, April.

International Seed Federation. *Seed Statistics*, Nyon, Switzerland, available at: www.worldseed.org/isf/seed (accessed November, 2010).

James, Clive. 1997. *Progress in Public-Private Sector Partnerships in International Agricultural Research and Development*, ISAAA Briefs No. 4-1997, International Service for the Acquisition of Agri-biotech Applications.

Kalton, R.R., and P.A. Richardson. 1983. "Private Sector Plant Breeding Programs: A Major Thrust in U.S. Agriculture," *Diversity*, November-December 16-18.

Kalton, R.R., P.A. Richardson, and N.M. Frey. 1989. "Inputs in Private Sector Plant Breeding and Biotechnology Research Programs in the USA," *Diversity* 5, 4: 22-25.

Klotz, Cassandra, Keith O. Fuglie, and Carl E. Pray. 1995. *Private Sector Agricultural Research Expenditures in the United States, 1960-1992*, Staff Paper AGES-9525, U.S. Department of Agriculture, Economic Research Service.

KWS. 2008. *Annual Report 2007/08*, KWS SAAT AG, 37555 Einbeck, Germany.

Le Buanec, Bernard. 2007. "Evolution of the Seed Industry During the Past Three Decades," *Seed Testing International*, ISTA News Bulletin No. 134, October, pp. 6-10.

Le Buanec, Bernard. 2008. "Evolution of the Seed Industry During the Past 40 Years," *SEED News Magazine* XII, No. 4 (July/August).

Levin, Richard C., Wesley M. Cohen, and David C. Mowery. 1985. "R&D Appropriability, Opportunity, and Market Structure: New Evidence on Some Schumpeterian Hypotheses," *American Economic Review* 72, 2 (May): 20-24.

Limagrain. 2007. *Annual Report 2007*, 63720 Chappes, France. Monsanto. 2009. *Monsanto 2009 Annual Report*, St. Louis, MO.

Narrod, Clare A., and Keith O. Fuglie. 2000. "Private Investments in Livestock Breeding With Implications for Public Research Policy," *Agribusiness* 16: 457-470.

National Science Foundation (NSF). 2010. *National Patterns of R&D Resources: 2008 Data Update*, NSF 10-314, National Science Foundation, Division of Science Resources Statistics, available at: www.nsf.gov/ statistics/nsf10314/

National Science Foundation, Division of Science Resource Studies. *Research and Development in Industry*, (various issues).

Pardey, Philip G., Nienke Beintema, Steven Dehmer, and Stanley Wood. 2006. *Agricultural Research: A Growing Global Divide?* Food Policy Report, International Food Policy Research Institute.

Perrin, R.K., K.A. Hunnings, L.A. Ihnen. 1983. *Some Effects of the U.S. Plant Variety Protection Act of 1970*, Economic Research Report No. 46, North Carolina State University, Department of Economics and Business, August.

Pharmaceutical Research and Manufacturers of America (annual reports). Pharmaceutical Industry Profile, PhRMA, March.

Pray, Carl E., and Keith O. Fuglie. 2000. "The Private Sector and International Technology Transfer in Agriculture," in *Public-Private Collaboration in Agricultural Research: New Institutional Arrangements and Economic Implications*, K.O. Fuglie and D.E. Schimmelpfennig (eds.), Ames, IA: Iowa State University Press, pp. 245-68.

Pray, Carl E., and Keith O. Fuglie. 2001. "Policies for Private Agricultural Research in Asian Developing Countries," in *Tomorrow's Agriculture: Incentives, Institutions, Infrastructure and Innovations*, G.H. Peters and Prabhu Pingali (eds.), Hunts, England: Ashgate.

Pray, Carl E., and Keith O. Fuglie. 2001. *Private Investment in Agricultural Research and International Technology Transfer in Asia*, Agricultural Economic Report 805, U.S. Department of Agriculture, Economic Research Service, available at: www.ers.usda.gov/publications/ aer805/aer805/

Pray, Carl E., Keith O. Fuglie, and Daniel K.N. Johnson. 2007. "Private Agricultural Research," in *Handbook of Agricultural Economics, Vol. 3*, R.E. Evenson and Prabhu Pingali (eds.), Amsterdam: Elsevier, pp. 2607-2645.

Ruttan, Vernon W. 1982. *Agricultural Research Policy*, Minneapolis: University of Minnesota Press.

Scherer, F.M., and David Ross. 1990. *Industrial Market Structure and Economic Performance*, 3rd edition, Boston: Houghton Mifflin.

Syngenta. 2008. *Syngenta Financial Report 2007*, Syngenta International AG, Basel, Switzerland, February.

Syngenta. 2010. *Syngenta Financial Report 2009*, Syngenta International AG, Basel, Switzerland, February.

U.S. Department of Agriculture, (USDA). 2000. *Inventory of Agricultural Research*, Current Research Information System, National Institute for Food and Agricultural Research.

U.S. Department of Agriculture (USDA). 2007. *Inventory of Agricultural Research*, Current Research Information System.

U.S. Department of Agriculture, Economic Research Service (USDA/ ERS). Agricultural Research Funding in the Public and Private Sectors, Agricultural Research and Productivity Briefing Room, available at www. ers.usda.gov/data/agresearchfunding/

U.S. Department of Agriculture, National Agricultural Statistics Service (USDA/NASS). *Agricultural Prices* (various annual issues).

Wilcke, H.G., and J.L. Williamson. 1977. *A Survey of U.S. Agricultural Research by Private Industry*, Agricultural Research Institute.

World Bank. World Development Indicators Databank, World Bank, available at http://data.worldbank.org/data-catalog/world-development-indicators

PRIVATE RESEARCH AND DEVELOPMENT FOR CROP GENETIC IMPROVEMENT

Paul W. Heisey and Keith O. Fuglie

Seed[1] has been an essential input in crop production since the origins of agriculture, when farmers first began to save grains for replanting and to select seeds for desirable characteristics. The pace of crop genetic improvement accelerated early in the 20th century with the development of applied genetics and associated changes in plant breeding, seed production, and seed marketing. The development of the modern seed industry began about this time.

From the mid-1990s to the present, the private-sector seed industry has probably undergone more structural change than any of the other agricultural input industries covered in this study. Technological innovation in the form of modern, DNA-level biotechnology and changes in intellectual property rules have enabled private-sector companies to capture more value from the new seeds they develop. In the late 1990s, research intensity (R&D spending as a percentage of sales) in the seed industry accelerated past intensity in all other agricultural input sectors, and seed remains the most research-intensive sector to date.

Global Market for Crop Seed

Seed by Sector of Origin

Seed used for crops has three main sources: farmer-saved or farmer-sourced seed, commercial seed from the public sector, and commercial seed from the private sector. Private-sector proprietary seed dominates markets globally today, particularly in high-income countries.

The dominance of private companies as the primary source of crop seed is a relatively recent phenomenon. Historically, farmers saved seed from their own crops or obtained seed from their neighbors for replanting in the next season. In some cases, poor crops or other negative factors may lead to consumption of saved seed, and after such periods, farmers may purchase seed from food markets to replenish their supplies. This is particularly common in some developing countries and for certain crops. Farmer-saved or "bin-run" seed continues to be a seed source, even for a major crop such as wheat, in high-income agricultural economies such as the United States. Plant variety protection laws, however, tend to restrict or forbid practices such as sales to other farmers.

As scientific plant breeding developed, public-sector breeders were often the major sources of new crop varieties. Public-sector varieties have sometimes been multiplied and sold to farmers by private seed distributors, but these distributors may not conduct seed-related research. In some developing countries, government-owned companies also distribute public seed varieties. As the seed industry develops further, however, private companies that perform their own plant breeding and seed research can become increasingly important (Morris et al., 1998). In high-income countries, where private-sector research on crop improvement is significant, public and private research efforts focus on complementary, but different, stages of the research process. For example, in plant breeding in the United States today, most public R&D is oriented toward basic breeding methods or basic germplasm enhancement, while most private R&D is devoted to commercial cultivar development (Frey, 1996). This balance between public and private-sector roles has continued to evolve over time, particularly with the more widespread commercial use of biotechnology techniques (Traxler, 1999). For some commodities (like wheat and potatoes in the United States), the public sector continues to provide most of the finished varieties because of a lack of private-sector interest (Heisey et al., 2001). Both public- and private-sector seed from scientific crop improvement programs are considered commercial seed, but in this study, we distinguish private-sector proprietary varieties from seed originating in the public sector.

Estimates of the size of the global commercial seed market in 2006 vary between $20 billion and $34 billion. Estimates of commercial seed value based on sales by companies that develop the seed may be somewhat lower than estimates based on farmer purchases. For purposes of historical comparison and disaggregation, we take an intermediate sales-based figure for commercial seed reported by the Context Network (2007), $22.9 billion. In addition, the value of farmer-saved seed in 2006 is estimated at $6.1 billion. Based on these two amounts, the total value of crop seed used in 2006 is estimated at $29 billion.[2] The $22.9 billion of commercial seed can be further subdivided into $19.6 billion of proprietary seed ($11.8 billion of conventional proprietary seed, $7.8 billion of GM proprietary seed, and $3.3 billion of public-sector seed) (table 2.1).

Table 2.1 shows estimates of the real value (in constant 2006 U.S. dollars) of the world seed market. Between 1995 and 2006, real market sales of commercial public sector seed declined, while sales of proprietary seed increased markedly. In all likelihood, this trend began well before 1995. In recent years, market sales of proprietary seed have been at least six times those of seed from the public sector.

The first significant commercial sales of proprietary seed with GM traits occurred in 1995. Market sales of GM seed have increased rapidly; since 2006, they have exceeded 40 percent of the total sales value of proprietary seed. The real value of all seed has increased over the last 5 years, and this increase has been driven primarily by the expansion of the market for GM seed.

Table 2.1. Size of the global seed market

Year	Proprietary conventional seed	Proprietary genetically modified seed	Total proprietary seed	Public commercial seed	Farmer-saved seed	Total value of all seed	ISF value of total seed
			Million constant 2006 U.S. dollars				
1995	13,447	95	13,542	5,550	6,333	25,425	
2001	11,847	3,645	15,492	3,539	5,923	24,954	34,173
2002	11,210	4,148	15,358	3,483	6,390	25,231	33,631
2003	11,084	4,938	16,022	3,409	6,694	26,125	32,922
2004	11,525	5,869	17,394	3,315	6,616	27,325	32,013
2005	12,082	6,815	18,897	3,408	6,402	28,707	30,979
2006	11,800	7,800	19,600	3,300	6,100	29,000	34,000

Sources: USDA, Economic Research Service using Context Network (2007) for all columns except ISF value of total seed, which is from International Seed Federation (ISF). Values adjusted for inflation by the U.S. Gross Domestic Product implicit price deflator *(Economic Report of the President,* 2009).

Seed Markets by Region and Commodity

Over the past decade, commercial seed markets in three broad world regions— Asia-Pacific, Europe/Middle East/Africa, and North American Free Trade Association (NAFTA)— have been roughly equal in market sales, although amounts vary from year to year. Seed markets in each region have generally fluctuated around 30 percent of the global total, as calculated from data reported by the International Seed Federation (ISF). The seed market in Latin America/Caribbean makes up the remaining 10 percent. In more recent years, the Middle East/Africa portion—under 5 percent of the global total—can be separated from that for Europe. Industry analysts tend to estimate a higher percentage for NAFTA and a lower percentage for Asia-Pacific, as indicated by recent (2010) data from the ISF.[3]

For 2006, the Context Network estimated that field crops accounted for 77 percent of the global market for proprietary seeds. This may be further broken down into grains (46 percent), oilseeds (20 percent), and other field crops (11 percent). Vegetable and flower seeds made up 14 percent of the proprietary seed market, and forage and turf grass 9 percent. We assume that these percentages apply to all commercial seed, including public-sector seed, in the absence of data disaggregating public-sector seed.

The Context Network also estimated the value of seed markets for major field crops. Including "technology values" for GM traits and royalties, corn constituted about 25 percent of the total global market for commercial seed between 2001 and 2005. The next largest commodity in terms of seed market sales was soybeans, with over 12 percent of the total. Both wheat—the largest crop worldwide in terms of acreage—and cotton constituted about 4 percent each of the global commercial seed market. The value of the seed market for these two crops is roughly equivalent, even though world wheat area is over six times larger than world cotton area (a much higher proportion of annual cotton seed requirements is sourced through commercial markets, and cotton seed unit prices are considerably higher in part because GM traits have been incorporated into cotton but not wheat). Rice, the second most widely grown crop worldwide, accounted for just over 1 percent of the commercial seed market (most annual rice seed requirements are sourced from farmer-saved seed). Apparent discrepancies between sales-based and farmer-purchasebased estimates may be larger for major crops such as wheat and rice, for which seed saving is widespread, than for such crops as corn, soybeans, or cotton, for which a larger percentage of seed used is purchased from seed companies. Furthermore, if the full market value of hybrid rice seed were accounted for, particularly for China where it is subsidized, the estimated rice share of global commercial seed might be higher.

The Market for Crop Genetic Traits

An alternative approach to partitioning the market, by GM traits, may be of economic interest but is more difficult to undertake with available data. Both large multinational companies that conduct research in agricultural biotechnology and small biotechnology firms develop products in two broad classes—GM traits that may be inserted into crop seed or research tools that facilitate biotechnology procedures, such as gene discovery or genetic modification. Determining licensing revenue that these firms obtain from traits or tools is generally not possible directly.[4] Large multinational companies include licensing revenue within their broadly defined "seed" category and will not separate it from actual sales of seed. Neither overall sales nor specific revenue from licensing are available for most small agricultural biotechnology firms, and in general, it is not possible to estimate licensing revenues for research tools.

The market value of technology fees and royalties, however, may provide a rough estimate of the value of GM traits, although it is not identical to this value. For some of the major crops, technology fees and royalties represent a significant proportion of the cost of seed. Globally, by 2005, these fees and royalties constituted about half the total market value of seed in the case of cotton and about a quarter in the case of soybeans. These are crops for which significant portions of world area (over 40 percent for cotton, and over 60 percent for soybeans) are planted to GM seeds (see James, 2007, for an estimate of the distribution of GM crops worldwide). By 2005, only about 10 percent of the value of corn seed came from technology fees and royalties. Historically, private-sector firms have been able to capture a greater return for breeding effort devoted to corn through the use of hybridization, which guarantees the purchase of seed every year (Morris et al., 1998). The use of GM corn, however, has increased rapidly in recent years (James, 2007), and so the percentage of corn seed value obtained from technology fees and royalties is still rising.

Public-sector wheat seed breeding has remained relatively more important worldwide than public-sector breeding in some other major crops because successful wheat hybrids have not been widely deployed and because other technical factors have not allowed private breeders to capture a greater share of the returns from wheat breeding. Nonetheless, royalties have made up about an eighth of total wheat seed values in recent years. Particularly in Europe, and more recently in other high-income countries, institutional arrangements have evolved in which seed distributors pay royalties from the sale of wheat seed to the breeders, private or public, of the varieties involved. Because GM wheat has not yet been planted commercially, royalties from the sale of wheat seed do not reflect the value of GM traits for this crop.

For the largest national seed market, the United States, it is possible to estimate the value of GM traits in corn, soybeans, and cotton using price data on "biotech" and "nonbiotech" seed from the USDA's National Agricultural Statistics Service (USDA/NASS),[5]

USDA/NASS data on total crop area planted[6] and the share of total area planted to crops with GM traits,[7]and seeding rates from USDA's Agricultural Resource Management Survey (ARMS).[8] Using these data, we first calculated the value of germplasm for all seed used in a given crop, biotech and nonbiotech, by using the nonbiotech seed price. We then estimated the value of GM traits by applying the difference between the biotech and the nonbiotech price to the area planted to GM crops.[9] The estimates followed the same pattern as the global estimates, with trait value accounting for the greatest percentage of total seed value for cotton and the lowest percentage for corn. These differences among crops may reflect the degree of adoption of GM traits, the market value of GM traits in comparison to the market value of the underlying germplasm, or other market or technical factors that are less well understood. Biotech traits accounted for about half the total value of cotton seed in the United States in 2001. After continuing to rise for several more years, the share has remained at about 80 percent since 2007. Biotech traits represented only 7 percent of the U.S. corn seed market value in 2001 but rose to 28 percent in 2007 and to 37 percent in 2009. Biotech traits have fluctuated between 30 and 40 percent of the U.S. soybean seed market for much of the past decade, reaching a high of 42 percent in 2007 but falling to 30 percent in 2009. Based on these data, the value of biotech traits for corn, soybeans, and cotton taken together have represented 20 percent or more of the value of the entire U.S. seed market in recent years.

Several studies in recent years used national micro-level market data to analyze component pricing strategy for different biotechnology traits and market power for U.S. field crops for which GM varieties are important. For the case of corn (Shi et al., 2010a) and soybeans (Shi et al., 2009), these data can also be used as a check on the estimates based on the NASS data.[10] For cotton, Shi et al. (2010b) do not present data on national-level seed prices, but their data still allow for comparisons with the NASS-based estimates used in this study.[11] Nonetheless, calculations based on data from Shi et al. suggest the same relationships (e.g., the proportion of the value of the U.S. corn seed market attributable to GM traits is lowest among the three major crops with significant areas planted to GM varieties, and the proportion of the value of the U.S. cotton seed market made up of GM traits is the highest). Similar to our NASS-based calculations, calculations based on Shi et al.'s (2010a) data imply the trait percentage of value for corn has risen from 7 to 21 percent between 2001 and 2007, while the trait percentage of value for soybeans has been higher but more variable, estimated at around 35 percent in 2007.

Structure of the Global Seed Industry

Seed Company Market Structure in 2009

In this report, we divide the global crop seed and biotechnology industry into four tiers. The first tier is what we refer to as the "Big 6"—large multinational corporations with positions in both the markets for crop seed-biotechnology and agricultural chemicals. The second tier consists of other seed companies that do some research. This group includes companies that may have a significant global presence as well as smaller regional or local seed companies. A third tier consists of mostly small seed companies that do not conduct research themselves but only produce and sell seed under licensing or other commercial arrangements with the other companies or public-sector breeders who develop new varieties. Finally, a fourth tier comprises small and medium-size agricultural biotechnology companies.

These companies do not generally sell seed but rather seek to commercialize a new genetic trait or biotechnology service or tool to other firms in the industry.

Table 2.2. Companies with over $100 million in crop seed and biotechnology sales in 2009, plus BASF

Company	Country of incorporation	Crop seed and biotech sales	Agricultural chemical sales	Nonagricultural chemical sales and R&D	Pharmaceutical sales and R&D	Agricultural biotechnology research
		Million U.S. dollars				
Monsanto	U.S.	7,297	3,527	Divested 1997	Divested 2000	>80% of crop R&D
DuPont/Pioneer	U.S.	4,806	2,320	Primary product	Divested 2001	>50% of crop R&D
Syngenta	Switzerland	2,564	8,491	Divested 1996	Divested 2000	>15% of crop R&D
Limagrain	France	1,370	0	--	--	>25% of crop R&D
KWS AG	Germany	996	0	--	--	Yes
Bayer	Germany	699	7,535	No	Human and animal health	>85% of crop R&D
Dow	U.S.	633	3,708	Primary product	Divested 1996	>85% of crop R&D
Sakata	Japan	485	0	--	--	Yes
Forage Genetics Int'l (Land O'Lakes)	U.S.	*412*	0	--	--	No
DLF-Trifolium	Denmark	*391*	0	--	--	Yes
Takii	Japan	*347*	0	--	--	Yes
Rijk Zwaan	Netherlands	265	0	--	--	Yes
In Vivo	France	217	0	--	--	Yes
BarenBrug Holland BV	Netherlands	208	0	--	--	Yes
Saaten-Union	Germany	*187*	0	--	--	Yes
RAGT Semences SA	France	*181*	0	--	--	Yes
Florimond Desprez	France	*162*	0	--	--	Yes
Euralis Group	France	154	0	--	--	Yes
Maisadour Semences	France	119	0	--	--	Yes
Stine Seeds	U.S.	*unknown*	0	--	--	Yes
BASF	Germany	*small*	5,065	Primary product	Divested 2000	100% of crop R&D

* Seed sales figures for Land O'Lakes refer to alfalfa/forage seed developed by Forage Genetics International. Land O'Lakes also distributes seed for other companies, such as Monsanto and Syngenta, but these sales are not included in the Land O'Lakes estimate.

Seed sales figures in italics are ERS estimates not derived directly from company data.

Sources: USDA, Economic Research Service using compiled company reports and press releases, Le Buanec (2007), Allison (2007), and PhillipsMcDougall.

Our survey identified more than 100 such "agricultural biotechnology startups" that have proliferated since the 1980s. Some of these companies have been bought out by larger companies and others have exited the market, but at least 30 startups were in operation as of 2008. Table 2.2 summarizes information on 21 companies that each had global seed sales of over $100 million in 2009, including BASF, which made significant investments in crop biotechnology but did not have substantial seed or trait sales. Six of these companies also had a significant market share for crop protection chemicals.

The Big 6 firms (Monsanto, Dupont, Syngenta, Bayer, Dow, and BASF) hold a unique position in integrating biological and chemical technologies in agricultural input markets. In 2009, three of these firms—Monsanto, DuPont/Pioneer, and Syngenta—were the top three global seed companies, and they also ranked fifth, sixth, and first, respectively, in global sales of crop protection chemicals.

Bayer, another Big 6 firm, only entered the seed market in 2002 with the acquisition of Aventis Crop Science; by 2009, it ranked sixth in global seed sales. Dow (fourth in crop protection sales in 2009) ranked seventh in global seed sales, and BASF (third in global crop protection sales), which began research in seed/biotechnology around 1998, reported no seed or trait sales as of 2009.[12] However, both BASF and Dow have made significant investments in agricultural biotechnology.

Two European-based companies, Limagrain and KWS, also had seed sales of over $600 million in 2009. Limagrain, originally a producer-owned cooperative based in France, is active in both the markets for seeds of field crops and vegetables. KWS, based in Germany, concentrates on seeds of field crops. Limagrain and KWS have a joint venture, AgReliant, in the North American market. In 2009, at least 13 companies had global seed sales between $100 million and $600 million. Three companies (Land O'Lakes/Forage Genetics International, DLF-Trifolium, and BarenBrug Holland BV) specialized in seed of forage and/or turf grass. Another Dutch seed company, Rijk Zwaan, and two Japanese-based companies, Sakata and Takii, specialized in vegetable and/or flower seed. Several European-based companies, In Vivo, Saaten-Union, RAGT, Florimond Desprez, Euralis Group, and Maisador Semences, focused mainly on field crops but also had forage/turf grass products as well as sugar beets. Finally, the U.S. soybean breeding company, Stine Seeds, likely fits into this category as well, although Stine Seeds does not disclose sales information. In 2009, these companies (except BASF) invested in crop improvement and sold proprietary seed for a range of crops (fig. 2.1).

Changes in the Structure of the Global Seed Industry

Over the past 15 years, the seed industry has consolidated through mergers and acquisitions (see fig. 2.2 for global activity; Fernandez-Cornejo (2004) for U.S. activity; and Howard (2009) for graphics on current and historical seed industry ownership, including that of many small companies.[13]

Some of the features of changes to the seed industry can be summarized as follows:

1) Among the largest firms in terms of total product sales, the *close relationships between seed and agricultural chemicals industries* have continued. This applies to the Big 6 firms in particular (see fig. 2.2). These relationships may result partially from complementarity of product lines such as herbicide-tolerant seeds and chemical herbicides (Just and Hueth, 1993), or possibly from economies of scope in marketing as well.

   Chemical companies also realized GM crops with pest resistance traits would compete with the crop protection chemicals, which helped drive these companies' interest first in biotechnology and eventually in seed, thus changing their business models to meet farmer demand for crop pest management as technological opportunities changed.

2) On the other hand, the *"life science industry" model* suggested a decade ago (Enriquez, 1998) *has not become the dominant paradigm*. This model stemmed from the likelihood that technologies underlying pharmaceutical discovery were the same as those underlying gene discovery for seeds. Differences in business models and types of customer, however, prevented firms from combining both pharmaceuticals and agricultural biotechnology. Of the current Big 6 companies, only one— Bayer— has pharmaceuticals as its primary product line. Even when Bayer expanded into the seed/biotechnology industry in 2002 with its acquisition of Aventis Crop Science, Aventis pharmaceuticals eventually became a component of Sanofi-Aventis pharmaceuticals, not Bayer. Monsanto, which entered pharmaceuticals in the mid-1980s with its acquisition of Searle, was briefly held by Pharmacia before the agricultural enterprise was spun off as the "new Monsanto"; Pharmacia retained the pharmaceutical business segments. When Novartis's chemical and seed businesses were merged with Zeneca's agricultural chemical business in 2000 to form Syngenta, the pharmaceutical portion of Novartis remained intact as a separate large pharmaceutical company. BASF and DuPont ended their relatively smaller pharmaceutical investments after 2000 and 2001, respectively, and Dow had already sold its pharmaceutical subsidiary Marion Merrell Dow to Hoechst in 1996.

3) Agricultural chemicals have been an important part of product sales for all the Big 6 companies. However, positions in markets for nonagricultural chemicals have not remained constant, with some companies shedding these nonagricultural products. Monsanto divested this portion of its business to Solutia in 1997.
   When Ciba-Geigy and Sandoz merged to form Novartis in 1996, nonagricultural chemicals were spun off to Ciba Specialty Chemicals, which eventually was acquired by BASF.
   In response to antitrust considerations, Bayer sold selected insecticides and fungicides to BASF in 2003. DuPont sold its polymers business in the early 2000s.

Despite the common features in structural changes in the current Big 6, these and other former large multinational seed companies have followed somewhat individualized trajectories. Three of the current Big 6 were already identified as industry leaders 7 to 10 years ago: Monsanto, DuPont/Pioneer, and Syngenta (or Novartis) (Shoemaker et al., 2001; Fernandez-Cornejo, 2004). But several other potentially large players a decade ago no longer exist. Aventis Crop Science was acquired by Bayer in 2002, giving Bayer a position in the seed/biotechnology industry. Astra-Zeneca's seed business became part of the Advanta Seed Group, but Advanta's seed enterprise was broken up with parts acquired by Syngenta, Limagrain, and others in 2004 and 2005.

Monsanto has transformed itself most completely, from a chemical company to a seed/biotechnology company, and made by far the greatest number of large acquisitions of seed and related companies. Although Monsanto still has significant sales of the herbicide glyphosate, its research investments in chemicals are markedly reduced. Syngenta maintains a strong position in crop protection as well as in seed, but in contrast to its legacy companies, it is a wholly agricultural company.

Company	Country	Maize	Wheat & small grains	Rice	Soybean	Canola/ rape/sun-flower	Cotton	Sugar-beet	Sugar-cane	Forage	Vegetables & flowers
Monsanto	U.S.	Y	Y		Y	Y	Y	*	Y	*	Y
DuPont/Pioneer	U.S.	Y	Y	Y	Y	Y	Y			Y	
Syngenta	Switzerland	Y	Y	*	Y	Y	*	Y			Y
Limagrain	France	Y	Y	Y	Y	Y				Y	Y
KWS AG	Germany	Y	Y					Y			
Bayer	Germany			Y		Y	Y				Y
Dow	U.S.	Y	*	*	Y	Y	Y			Y	
In Vivo	France		Y			Y				Y	
Saaten-Union	Germany	Y	Y			Y				Y	
Florimond Desprez	France		Y			Y		Y		Y	
Stine Seeds	U.S.	Y			Y						
RAGT Semences SA	France	Y	Y		Y	Y				Y	
Euralis Group	France	Y				Y					
Maisadour Semences	France	Y				Y					
Forage Genetics Int'l	U.S.									Y	
DLF-Trifolium	Denmark									Y	
BarenBrug Holland BV	Netherlands									Y	
Takii	Japan										Y
Sakata	Japan										Y
Rijk Zwaan	Netherlands										Y

Y = company has crop breeding and seed sales; * = company develops biotechnology traits and platforms only. Source: USDA, Economic Research Service using data compiled from company websites.

Figure 2.1. Crop R&D portfolios of leading seed companies.

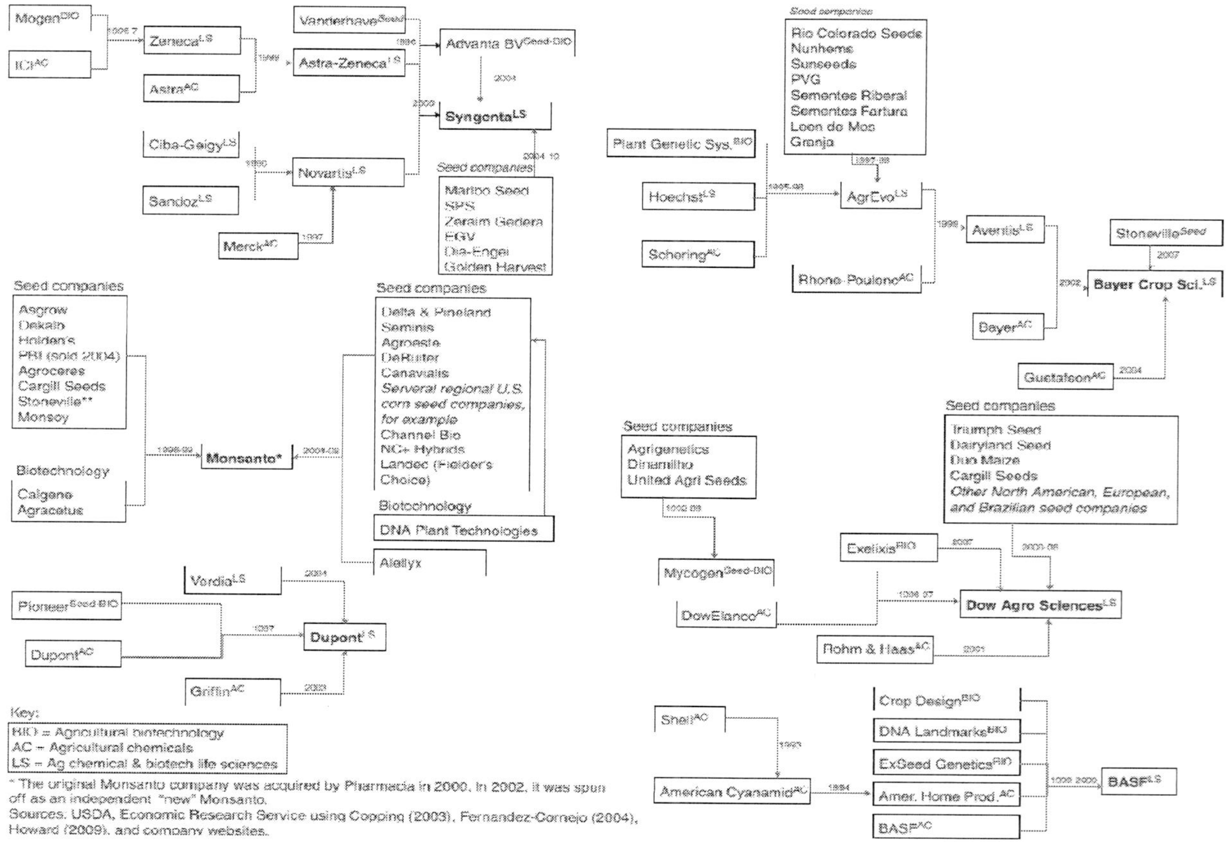

Figure 2.2. Formation of the "Big 6" seed-biotechnology-crop companies.

DuPont made by far the largest acquisition of all when it absorbed Pioneer, at the time the world's largest seed company, in 1999.[14] DuPont/Pioneer, however, has acquired only a few other seed or biotechnology firms since the merger. Dow was also recognized for its biotechnology investments a decade or more ago, particularly since it purchased Eli Lilly's share in Dow Elanco in 1997 and formed Dow Agrosciences. Though Dow Agrosciences has since acquired regional seed companies, a large portion of its investment in seed/biotechnology research has been in biotechnology. BASF has also focused almost completely on biotechnology, although for 10 years it had an alliance with the Swedish seed company Svalöf Weibull. Starting in 2007, BASF has concluded a number of research collaboration and licensing agreements with Monsanto and other seed companies for commercialization of future biotechnologies that BASF may develop. As noted, Bayer's entry into the seed business came with its acquisition of Aventis Crop Science. In general, as large chemical companies decided to commercialize their own biotechnology research or to buy seed company research, they needed to get access to seed companies either through direct acquisition, joint-venture, or licensing agreements. Through its purchase of Pioneer, DuPont gained immediate access to seed for multiple crops in multiple regions. Other companies had to acquire companies focusing on different crops in different regions; Monsanto has followed this strategy most comprehensively.

Many other large multinational seed companies have also followed a pattern of acquisitions or joint ventures. For example, Limagrain, which acquired other European seed companies starting with Vilmorin in 1975, acquired several more seed companies in the late 1990s and early 2000s, including Advanta Europe in 2005. Limagrain also formed a joint venture with KWS, AgReliant, in the North American corn and soybean markets, and also participated in projects with small biotechnology companies. DLF-Trifolium acquired the Dutch company Cebeco in 2002. Syngenta has held a significant minority interest in Maisadour for years. Recently, Svalöf Weibull (now Lantmännen SW Seed) and Florimond Desprez have initiated a research alliance, as have RAGT Semences and In Vivo.

Changes in Concentration in the Global Seed Market

In 2009, the top four companies accounted for 54 percent of the global commercial seed market (including public sector commercial seed), and the top eight companies accounted for 63 percent of total commercial seed sales (see table 1.7 in chapter 1). In 1994, these shares were 21 percent and 29 percent, respectively. Four-firm concentration ratios as measured by seed sales are now roughly equivalent across large agricultural input sectors. The eight-firm concentration ratio is still lower than eight-firm concentration in the agricultural chemicals industry (although four of the top six companies in each segment are the same), the animal health sector, and the animal genetics sector. Furthermore, the rate of growth of concentration in the seed market has been greater than the rate of growth for agricultural chemicals, animal health, or farm machinery.

The Herfindahl index (HI) is sometimes preferred to concentration ratios because it is sensitive to the distribution of market shares among firms, while the concentration ratios are not. In 1994, the HI for the global crop seed market was lower than similar indices for agricultural input sectors like animal health and agricultural chemicals. Between 1994 and 2009, HIs for seed, agricultural chemicals, animal health, and farm machinery all rose, but the rate of growth was most rapid for the seed sector. In 2009, the ranking for crop seed was higher than for any other sector reported in table 1.7, with the exception of animal genetics.

Different sectors of the seed market can be more concentrated than the aggregate global seed market. For example, we estimate that the top four companies in the commercial market for vegetable seeds covered 70 percent of the global market in 2007, and the top eight companies 94 percent. Three of the Big 6 seed-chemical companies are major players in the global market for vegetable seeds—Monsanto since its acquisition of Seminis in 2005, Syngenta since its absorption of parts of the Advanta Seed Group in 2004, and Bayer since its acquisition of Aventis/Nunhems in 2002. Limagrain also sells a substantial amount of vegetable seeds and seeds for home gardens. Together with Sakata, Takii, and Rijk Zwaan, these companies represent the largest current participants in the vegetable seeds market.

Similarly, concentration in seed for particular crops in important individual markets can be higher than global concentration overall. For example, the four largest companies accounted for an estimated 72 percent of the U.S. market for corn (maize) seed in 2007 and 55 percent of the U.S. market for soybean seed (NRC, 2010). These shares have most likely continued to rise.[15] Data from USDA/Agricultural Marketing Service suggest that in 2009, the top two cotton seed companies, Bayer (Fibermax and Stoneville) and Monsanto (Deltapine), had 85 percent of the U.S. market for cotton seed, and the top four companies had 95 percent. These recent estimates compare with four-firm concentration ratios in 1998 of 67 percent for corn, 50 percent for soybeans, and 95 percent for cotton (Fernandez-Cornejo, 2004).

Agricultural Biotechnology and the Seed Industry

The structure of the seed industry has been transformed in large part because of the advent of modern agricultural biotechnology. Under the broadest definition, domestication of plant species and selection of desired characteristics within agricultural species would qualify as biotechnology. More narrowly, what *biotechnology* represents today is new knowledge about the natural processes of DNA replication, breakage, ligation, and repair that has made possible a deeper understanding of the mechanics of cell biology and the hereditary process itself (McCouch, 2001). Over the last 20-30 years, the term "biotechnology" in agriculture has been most closely associated with *genetic engineering*, but it may refer to a variety of techniques and products, including the use of molecular markers in genetic improvement or more general use of genomic information.

Although these modern techniques may appear to be simply additions to the genetic improvement toolkit, from an economic perspective, genetic engineering in particular has meant the development of three complementary markets: the traditional market for improved germplasm (or seed), the market for genes conferring traits that can be used to capture value, and the market for platform technologies or research tools. Intellectual property has become more prominent as a means of protecting traits, tools, and, to a certain extent, germplasm. Much of the merger and acquisition activity within the crop seed industry, as well as some research alliances and licensing agreements, is motivated by the desire to obtain access to products from all three of these markets, which are necessary for the final product of GM seed (Graff et al., 2003; King and Schimmelpfennig, 2005; and Marco and Rausser, 2008). Kalaitzandonakes and Bjornson (1997) argue that more well-defined and stronger intellectual property rights would have encouraged greater use of contracting and licensing arrangements and thus reduced the level of mergers and acquisitions.

In addition to intellectual property and R&D costs, costs of regulatory approval for agricultural biotechnology products constitute another type of fixed cost that can create barriers to market entry and thus influence industry concentration (Fulton and Giannakis, 2002). Kalaitzandonakes et al. (2006) estimate compliance costs for guiding a single genetic engineering event, for either insect-resistant corn or herbicide-tolerant corn, through the regulatory process at between $6 million and $15 million. Heisey and Schimmelpfennig (2006) argue that through the mid-2000s, the combination of desired characteristics and development of marketing and distribution networks were stronger determinants of industry concentration than regulation. However, it is likely that longer regulatory delays (Kalaitzandonakes et al., 2006) and regulatory costs for generic firms[16] as GM traits come off patent (Just, 2006) will have increasing effects on industry structure. Furthermore, the relative impact of regulatory costs might vary across crops as well as across countries (Pray et al., 2006) and be significantly higher in many developing countries (Falck-Zepeda and Cohen, 2006).

Even given the increasing concentration of the seed industry with the advent of agricultural biotechnology, many different types of institutions may perform crop biotechnology research: public research institutions which may focus on more basic research (for example on research tools, in noncommercial model plants (e.g., *arabidopsis*) to understand gene function, or for crops which are not served by the private sector); small or medium-sized biotechnology companies that may or may not specialize in crop biotechnology; large seed companies; and integrated seed-chemical companies currently exemplified by the Big 6.

Small and Medium-Size Biotechnology Companies

Biotechnology companies specialize in research tools, identification and development of traits, or both. Many of these companies have been high-risk startups, depending on venture capital or "angel" investors, and turnover among these companies has been rapid. Significant entry into the marketplace began in the late 1970s and early 1980s, but in recent years, exits have outnumbered entrants. We identified 77 small and medium-sized companies[17] that have entered the agricultural biotechnology market since 1979 (fig. 2.3). All of these companies had agriculture as their primary business segment, and all but eight focused on crop biotechnology (these eight focused on animal biotechnology). Just over 30 companies are active as of 2008. Of the exits from the industry, about three-quarters were the result of acquisition and the remainder was due to bankruptcy or divestment.

In nearly all cases, we can only make estimates of the research investment by these firms, so the total estimated research expenditure by small biotechnology companies essentially provides the same information as a count of the number of active companies. In recent years, the total research investment of these companies is estimated to be about 5 percent or less of the total private-sector investment in seed/biotechnology research. Nonetheless, these companies developed some key agricultural biotechnology products, even though other companies eventually took the products to final market. For example, acquisitions of Agracetus and Calgene by Monsanto; Mogen by the Advanta group (now Syngenta); Mycogen by Dow; Plant Genetic Systems by AgrEvo (now Bayer, through Aventis); and DNA Plant Technologies by Seminis (now Monsanto) have all contributed significantly to the eventual holders' portfolios of traits and research tools (Graff et al., 2003).

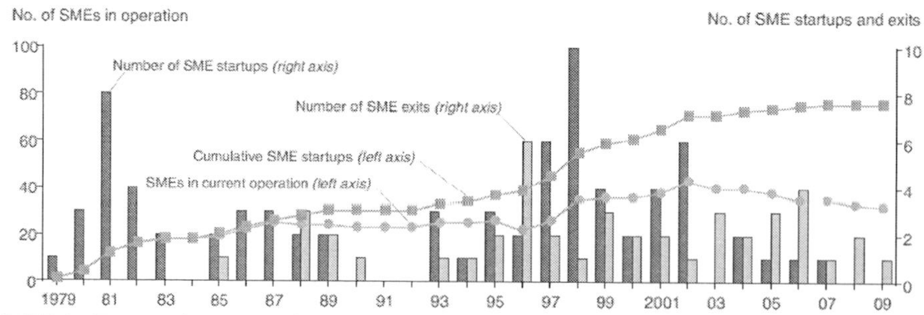

Source: USDA, Economic Research Service using data compiled from company websites and print media. Includes data on 77 small and medium enterprises (SMEs) for which agricultural biotechnology was or is the major focus.

Figure 2.3. Small and medium agricultural biotechnology startups and exits since 1979.

Large Seed Companies and the Big 6

All of the seed companies listed in table 2.2 have instituted biotechnology research, acquired interest in biotechnology companies, collaborated in biotechnology research, or signed licensing agreements for biotechnology products. In many cases, they have combined more than one of these activities. In several instances, seed companies that have not initiated GM research have still made use of marker-assisted breeding.[18] While the fixed cost of establishing inhouse capacity in biotechnology may be prohibitive for small or midsized seed companies, strategic partnerships between firms can enable such firms to access biotechnology. For example, a number of midsized vegetable breeding companies established a joint venture, Keygene, to conduct biotechnology research on their commodities of interest. Significant cross-licensing agreements still exist between companies, including those comprising the Big 6 (Howard, 2009). For example, Monsanto has cross-licensing agreements with all the other Big 6 companies; Dow with four of the other five, and DuPont and Syngenta with three of the other companies. Despite the increase in strategic partnerships and research collaboration, in terms of both current biotechnology research expenditures and current control of GM traits, only a few large companies dominate the market.

Trends in Private R&D Investment and Innovation in Crop Genetic Improvement

Research Spending by Private Seed and Crop Biotechnology Companies

Globally, real private sector research expenditures on crop seed development and crop biotechnology have risen substantially since the mid-1990s (table 2.3). This increase can be decomposed in a number of ways. First, research intensity for seed increased over much the 1990s. Since 2000, research intensity has fallen in the seed industry, although it is still higher than that for animal health, animal genetics, or agricultural chemicals, the other agricultural input industries with high research intensities. But over most years since 2000, the real value of the global seed market has increased (see table 2.1), and real research expenditures have continued to grow. The slight decline in research intensity in recent years has been the result of sales increasing even more rapidly than research investments (see tables 1.7 and 1.9 for data on research intensities across agricultural input industries).

Structurally, these increases in research expenditures have been marked by the remarkable rise in the share accounted for by integrated seed-chemical companies, characterized by the Big 6.[19] The share of total private-sector crop seed and biotechnology R&D spending by seed-chemical companies began to grow with the first round of mergers and acquisitions in the mid-1990s, and it surpassed the share of other large seed companies in 1999, the year that DuPont acquired Pioneer.

Table 2.3. Private-sector expenditures on crop seed and biotechnology research and development (R&D)

Year	Agricultural seed-chemical companies[1]	Other seed companies	Small and medium biotechnology firms	Total private seed and biotechnology R&D
		Million constant 2006 U.S. dollars		
1994	320	976	166	1,462
1995	355	1,013	168	1,536
1996	292	1,149	202	1,643
1997	576	1,139	144	1,859
1998	786	1,164	128	2,078
1999	1,091	908	128	2,127
2000	1,411	857	126	2,394
2001	1,312	842	141	2,295
2002	1,239	826	150	2,215
2003	1,266	857	142	2,265
2004	1,394	802	131	2,326
2005	1,453	746	129	2,328
2006	1,574	691	108	2,374
2007	1,764	676	100	2,540
2008	2,157	691	94	2,941
2009	2,353	702	93	3,149
2010	2,653	732	92	3,477

[1]These are companies with sales and R&D in both crop protection chemicals and seed/biotechnology. Since 2002 this group has been composed of the "Big 6" (BASF, Bayer, Dow, Dupont, Monsanto and Syngenta). Previously, AgrEvo, Astra-Zeneca, Aventis, Ciba-Geigy, Novartis and Sandoz were also part of this group.

Sources: USDA, Economic Research Service using data compiled from company reports, ASGROW (2007), Duncan (2007), author interviews with selected companies, and author extrapolations as described in text. R&D expenditures adjusted for inflation by the U.S. Gross Domestic Product implicit price deflator (*Economic Report of the President, 2009*).

In 2007, seed-chemical companies accounted for over 70 percent of total global R&D spending by the seed-biotechnology industry. Other seed companies made up an additional 24 percent, while small and midsized biotechnology firms accounted for 4 percent of R&D spending by this industry. The average research intensity of the largest seed companies is also higher than that of midsized and small seed companies, but small biotechnology startups as a group account for the highest research intensity of all, at over 40 percent in 2006 (see table 1.9).

Regionally, firms based in NAFTA countries, particularly the United States, and in Europe/Middle East/Africa (primarily Europe) dominate private research spending in crop seed and biotechnology. In 2006, over 53 percent of total investment was made by U.S.-based firms and another 42 percent was made by European-based firms. About 4 percent was attributed to Asia-Pacific firms and less than 1 percent to firms based in Latin America. With the globalization of the seed industry, however, many large firms are spending research dollars in other than their home regions.

It is not easy to estimate how much research is devoted to any particular crop. Companies breeding multiple crops may share resources, such as biotechnology-enabling technologies, across crops or may have other expenses, such as the costs of regulatory compliance, that are difficult to allocate. Nonetheless, it is clear that the largest proportion of seed/biotech research is directed toward corn (maize). Expert opinion also suggests that corn accounts for about 45 percent of all private-sector seed-related research (Cavalieri, 2009). This share is considerably higher than corn seed's 25-percent share of the overall seed market, but it is in accordance with industry estimates that corn seed is the most profitable seed for private companies to produce. Besides sales, profitability would be expected to contribute to relative seed research investment. We do not have enough information to estimate how much research is conducted on other individual crops, but we hypothesize that soybeans, cotton, and wheat research all command higher proportions of the private-sector total than their respective shares of the global seed market.

It is also difficult to estimate how much private R&D is devoted to conventional breeding research versus biotechnology. Le Buanec (2007) provides estimates of the relative proportions of seed-related research using biotechnology (including both genetic modification and marker-assisted selection) for several major companies in 2003. Using these estimates, and based on the assumptions that (1) 5 percent of other large seed companies' (see table 2.2) research is allocated to biotechnology, (2) 1 percent of other seed companies' research is allocated to biotechnology, and (3) all of the research for dedicated crop biotechnology companies is, indeed, allocated to biotechnology, we estimate that about half of the total research expenditures by the private seed industry in 2003 were allocated to biotechnology-related endeavors.[20]

Ownership Concentration in Crop Biotechnology Innovations

While data on research expenditures provide some indication of investments in innovation, they do not provide much information on the amount of new innovation actually taking place or its ownership. To gauge the sources of technological innovation emanating from private R&D, we assembled a number of indicators of intermediate research outputs and assigned them to the company or institution producing them. Our choice of indicators was necessarily limited by data availability. For example, we did not obtain counts of new crop varieties released (a reliable indicator of crop research output) due to the difficulty in compiling a relatively complete dataset with wide global coverage from publicly available sources.[21] The most complete data in this case might be that for European varietal registration, but linking named varieties with holders of plant variety protection rights would be extremely difficult. In the world's largest seed market, the United States, data are available for varieties with plant varietal protection certificates, varieties with utility patents, and varieties submitted for cultivar registration, but these data are both overlapping and nonuniversal in coverage.

The indicators reported here include the number of (1) agricultural biotechnology patents[22] issued in the United States, (2) petitions and notifications to USDA's Animal and Plant Health Inspection Service (USDA/APHIS) to import, transport across State lines, or release into the environment GM crops in the United States, (3) global approvals for planting or environmental release of GM crops, and (4) acres (multiplied by the number of traits per acre to get "trait-acres") planted to GM crops worldwide. These indicators are undoubtedly influenced by research investments, but the relationships are likely to be nonlinear and may change over time.

We summarize these data by showing the shares of select output held by a particular company or institution (see table 2.4). Those shares assigned to the eight listed seed companies include assignments to their historical counterparts, acquisitions (e.g., "legacy companies"), and subsidiaries. The counts for each indicator are cumulative (summed over several years) and therefore indicative of the overall dominance of the firm or institution in the market for seed and biotechnology innovations. But the indicators do not reflect changes in the relative market position of the firms or institutions over time. For example, patent counts are the total issued to a firm between 1982 and 2007 in the case of patents for crop cultivars, and between 1976 and 2000 for other classes of agricultural biotechnology patents (described below). The counts of USDA/APHIS petitions and notifications are the number issued since 1985, when USDA/APHIS first began issuing permits regulating the use and movement of GM seed, through mid-2008.[23]

Agricultural patents are counted three ways. The first column of table 2.4 is based on U.S. patents granted for crop cultivars through 2007. Three firms accounted for nearly three-fourths of these patents; these operations are the three largest seed companies in both the United States and the world. Ninety percent of these patents were issued for two crops: corn (either inbred parent lines or hybrids) and soybeans, with cotton a distant third with just over 2 percent of total patents. One midsized U.S. seed firm, Stine Seeds (not listed separately in the table), held nearly 13 percent of the total soybean cultivar patents. As might be expected, public or nonprofit institutions held few cultivar patents.

The second and third columns in the table summarize data from ERS's Agricultural Biotechnology Intellectual Property (ABIP) database. The ABIP database includes all agriculturally related biotechnology patents granted between 1976 and 2000. The data in the second column use a narrow definition for agricultural biotechnology, namely, patents that pertained specifically to crops and to the use of "modern" biotechnological techniques such as genetic engineering, selection with the aid of molecular markers, or genomics. The data in the third column are based on a broader definition of agricultural biotechnology, where biotechnology refers to any "use of organisms or parts of an organism to make or improve products or processes" in agriculture or food production. Both the narrow and broad patent definitions include patents issued for crop cultivars (the patents counted in the first column) through 2000. The Big 6 companies account for nearly two-thirds of the modern crop biotechnology patents (narrow definition) and one-quarter of the more broadly defined agricultural biotechnology patents. As with cultivar patents, the majority of these patents were issued in the names of legacy companies or subsidiaries. Unlike with cultivar patents, nonprofit institutions hold a notable minority of these biotechnology patents. Unfortunately, categorizing biotechnology patents is difficult and subject to error, and data coverage extends only through the end of the year 2000. Thus, patenting by more recent entrants may be underreported.

The fourth column presents data on petitions and notifications to USDA/ APHIS on the importation, interstate transport, and environmental releases of GM seed. The Big 6 firms accounted for 62 percent of these petitions and notifications, with Monsanto claiming nearly 40 percent of the total.

Table 2.4. Shares of selected research outputs held by major seed companies and other institutions

Company (including subsidiaries and acquisitions)	U.S. patents issued for crop cultivars	U.S. patents issued for agricultural biotechnology, narrowly defined[1]	U.S. patents issued for agricultural biotechnology, broadly defined[1]	Petitions and notifications to APHIS for field trials with GM plants in the U.S.	Approvals for planting or environmental release of GM crops	GM trait-acres of soybeans, corn, cotton, canola[2]
	% of total issued 1982 to 2007	% of total issued 1976 to 2000	% of total issued 1976 to 2001	% of total field trials 1985 to mid-2008	% of global approvals 1985 to 2007	% of global trait-acres in 2007
Monsanto	28.7	16.8	6.1	39.7	49.2	about 85
Dupont	36.5	20.7	5.1	7.3	5.0	3-5
Syngenta	8.4	9.8	3.8	5.2	8.0	3-5
Bayer	0.2	5.8	4.7	4.9	20.9	3-5
Dow	2.3	9.9	2.1	3.5	4.0	3-5
BASF	0.0	0.9	2.1	1.0	0.0	0
Limagrain	1.2	0.5	n.a.	1.2	0	0
KWS	0.4	0.1	n.a.	1	<0.1	0
Other private firms or	21.7	15.1	43.8	9.8		
individuals					12.9	2
Public and nonprofit institutions	0.6	20.4	32.3	26.4		
Total %	100.0	100.0	100.0	100.0	100.0	100.0
Total number				38,978	306	352 (282)[3]
						(mil. acres)

n.a. = not available. GM = genetically modified crops.

[1]The ERS Agricultural Biotechnology Intellectual Property (ABIP) database defines agricultural biotechnology two ways. Under a narrow definition, it includes patents pertaining specifically to crops and to the suite of modern biotechnology techniques, such as genetic engineering, selection with the aid of molecular markers, or genomics. Under the broader definition, biotechnology refers to any "use of organisms or parts of an organism to make or improve products or processes" in food or agriculture. See King and Schimmelpfennig (2005) for further information regarding ABIP.

[2] A "trait-acre" is the area sown to GM crops, where stacked GM traits are counted as multiple acres, depending on the number of traits stacked in a single seed. Total trait-acres include data from 13 countries with at least 100,000 ha in total GM crop area in 2007.

[3]The first figure is for total GM trait-acres. The second figure is the total area planted to GM crops in 2007 (James, 2007).

Sources: USDA, Economic Research Service using the following: U.S. patents issued from crop cultivars from the U.S. Patent and Trademarks Office; U.S. patents issued for agricultural biotechnology (narrow and broad definitions) from the ERS Agricultural Intellectual Property database; petitions and notifications to APHIS are from Virginia Polytechnic Institute and State University; approvals for plantings or environmental release of GM crops and GM trait-acres based on data provided in James (2007).

In this instance, more petitions were recorded in the name of the parent firm than in the names of legacy companies. Public and nonprofit institutions accounted for about one-quarter of these petitions.

The final two columns are based on data on global use of GM seed reported by James (2007). The second-to-last column summarizes approvals for planting and/or environmental release of a specific GM "event" through 2007.[24] These data are a simple count of the number of approvals granted in any country—if the same GM event has been approved in two countries, it is counted twice. The Big 6 firms accounted for 87 percent of these approvals, with Monsanto claiming nearly half of the total.[25] Outside of the Big 6, the most notable holder of approvals for GM events was Florigene, an Australian company with traits for altering flower color.

In the last column, we use data on the area planted to GM cultivars and event approvals from James (2007), together with other information, to estimate the total "trait-acres" under GM crops and assign them to a given source. A trait-acre consists of a single trait planted on an acre, which means that acreage with stacked genes (e.g., Bt and herbicide tolerance combined in a single cultivar) is counted more than once, depending on the total number of biotechnology traits that are "stacked' in a variety. This effect complicates the estimation, particularly when multiple stacked traits (which may originate from different sources) are widely deployed. We made these assessments for the top four GM crops (soybeans, corn, cotton, and canola) for the 13 countries worldwide in which more than 100,000 hectares of GM crops were planted in 2007.

Some individual trait-crop combinations are easier to assess than others. For example, Roundup Ready soybeans, which account for over 40 percent of all trait-acres, are easy to identify from the data in James (2007). On the other hand, insect-resistance traits are very hard to separate, particularly in corn.[26] To quote an Iowa State University extension publication on integrated pest management, the earliest Bt corn featured "one gene (Cry1Ab), three genetic events (176, Bt-11, MON 810) with event 176 under two trademarks (KnockOut and NatureGard) and events Bt-11 and MON 810 under one trademark (YieldGard)."[27] Furthermore, these Bt events were associated with a number of different companies: Monsanto; Northrup King/Sandoz/Novartis/ Syngenta; Ciba-Geigy/Novartis/Syngenta; and Mycogen/Dow.[28]

Based on the level of actual planting of GM crops in 2007, Monsanto traits clearly dominated, with approximately 85 percent of total global trait acres. Four of the other five Big 6 companies had some planted trait acreages, although it is difficult to estimate exact areas. Bayer was notable in that it held the most widely used events for resistance to glufosinate, a herbicide alternative to glyphosate. Glufosinate-tolerant ("Liberty Link") canola varieties were deployed early in Canada and have achieved a minor but notable position in both Canadian and U.S. canola production. Early glufosinate-tolerant events can be traced to Plant Genetic Systems, a biotechnology firm that was acquired first by AgrEvo, then became part of Aventis and eventually Bayer CropScience. Glufosinate tolerance is also being deployed in corn, cotton, and soybeans, but market share up to 2007 appeared to be relatively small for these crops. The only entity outside of the Big 6 companies with notable trait acreage in 2007 was the Chinese Academy of Agricultural Sciences, a public institution, with measureable contributions to Bt cotton planted in both China and India.

Patent data are probably more useful for testing hypotheses on the structure of the seed-biotechnology industry than they are as indictors for the amount of research investment. The amount of patenting in an industry depends on other factors in addition to research spending, such as changes in intellectual property policy (the advent of biotechnology changed and expanded what is considered patentable material in crops). Agricultural biotechnology patent data have been used to test a number of hypotheses about market structure, especially the relationship between intellectual property and mergers, acquisitions, and divestitures. This research has shown that these changes to market structure have been motivated by a need to combine complementary technology assets such as core germplasm, GM traits, and research tools (Graff et al., 2003); the desire to acquire greater depth within a particular technological area (King and Schimmelpfennig, 2005); and the relative enforceability of patents (Kalaitzandonakes and Bjornson, 1997; Marco and Rausser, 2008).

Patent and field trial data can also provide evidence on whether or not new entrants are a significant source of new innovations (Brennan et al., 2005). The USDA/APHIS field trial data in table 2.4 show more concentration than the biotechnology patent data, but Brennan et al. find that concentration in both patents and field trials has increased over time. They claim that new firm entry has increased, but concentration at the top of the industry has also increased. By some measures, U.S. cultivar patents are even more concentrated than other indicators, and at the global level, actually planted GM traits are dominated by a single firm. These concentration indicators may change as other major seed-biotechnology companies seek to commercialize new varieties and traits. Furthermore, tracking traits will become increasingly complex as multiple GM traits from a variety of firms are inserted into individual varieties.

Factors Influencing Future Trends in Private-Sector Seed-Biotechnology R&D

Structural change in the seed-biotechnology industry may have been greater than in any of the other industries covered in this study. Several factors will continue to influence this structure and the level and composition of private-sector investments in seed-biotechnology R&D:

1) *Potential for future market expansion.* Future market expansion will be determined by, among other things, the potential for greater use of improved seed in general and GM crops in particular in developing countries; by changes in consumer attitudes toward genetic engineering, particularly in high-income countries; by the potential for expansion in biotechnology applications to additional crops; and by the development of newer biotechnology applications, for example, tolerance to drought stress.
2) *Industry structure, seed pricing, and seed-biotechnology R&D.* Considerable attention has been focused on the effects of industry structure on seed pricing and potential distributional effects. For example, Stiegert et al. (2010) summarize a number of studies and find that in the United States, own-market concentration can increase seed prices. At the same time, the study finds that if a more integrated system of production of GM seeds by a few large firms reduces development costs, the resulting price effects may reduce or reverse price increases due to market power. Focusing specifically on the impacts of industry structure on R&D, Schimmelpfennig

et al. (2004), using field-trial data from the United States, argue that increases in industry concentration had a negative effect on research intensity in agricultural biotechnology. The evidence reported here on actual R&D expenditures for the global seed industry as a whole shows that the net trend in research intensity was strongly positive during the 1990s, and real private-sector research expenditures have continued to grow since then (see table 2.3). Alternatively, Kalaitzandonakes et al. (2010) show that the value of price premiums and markups for GM seed in the U.S. corn and soybean seed industries did not exceed R&D expenditures until 2007, and the authors cite this as evidence of dynamic efficiency in these industries.

3) *Interaction between other policies and seed-biotechnology R&D.* Concerns about potential anti-competitive behavior in the seed-biotechnology sector have expanded beyond distributional effects of seed pricing to questions of whether antitrust policy needs to consider R&D concentration and behavior in trait markets as well as in ultimate seed markets (Moss, 2009; Monsanto, 2009). Regulatory policy as some of the first GM traits (e.g., first generation herbicide tolerance) come off-patent may also determine the potential for entry by firms marketing generic GM crops. Small generic firms that might otherwise be willing and able to develop and market GM seed with an off-patent trait might be less able to meet regulatory compliance requirements than the larger firms that initially marketed the technology when it was still under patent.

References

AGROW Reports. 2007. *GM Crops*, Informa UK Ltd., London, July.

AGROW Reports. 2006. *Global Seed Industry*, Informa UK Ltd., London, September.

Brennan, Margaret, Carl Pray, Anwar Naseem, and James F. Oehmke. 2005. "An Innovation Market Approach to Analyzing Impacts of Mergers and Acquisitions in the Plant Biotechnology Industry," *AgBioForum* 8 (2&3): 89-99.

Cahoon, Richard S. 2007. "Licensing Agreements in Agricultural Biotechnology," in A. Krattiger, R.T. Mahoney, L. Nelsen et al. (eds.), *Intellectual Property Management in Health and Agricultural Innovation: A Handbook of Best Practices*, Oxford, UK: MIHR, and Davis, CA: PIPRA, pp. 1009-1016, available at: www.ipHandbook.org.

Cavalieri, Anthony. 2011. Former Vice-President and Director of Trait and Technology Development, Pioneer Hi-Bred International, 2009, personal communication.

Context Network. 2007. "Sizing Up the Seed Market," *Seed World*, pp. 10-12.

Copping, Len. 2003. "The Evolution of Crop Protection Companies," *Pesticide Outlook* 14 (December): 276-78.

Economic Report of the President. 2009. U.S. Government Printing Office.

Enriquez, Juan. 1998. *Gene Research, The Mapping of Life, and the Global Economy*, Harvard Business School Case 599-016, Cambridge, MA.

Falck Zepeda, José, and Joel I. Cohen. 2006. Biosafety Regulation of Genetically Modified Orphan Crops in Developing Countries: A Way Forward," in Richard E. Just, Julian M. Alston, and David Zilberman (eds.), *Regulating Agricultural Biotechnology: Economics and Policy*, New York: Springer.

Fernandez-Cornejo, Jorge. 2004. *The Seed Industry in U.S. Agriculture: An Exploration of Data and Information on Crop Seed Markets, Regulation, Industry Structure, and Research and Development*, Agriculture Information Bulletin Number 786, U.S. Department of Agriculture, Economic Research Service, available at: www.ers.usda.gov/publications/ aib786/

Frey, Kenneth J. 1996. *National Plant Breeding Study-I: Human and Financial Resources Devoted to Plant Breeding Research and Development in the United States in 1994*, Special Report 98, Iowa Agricultural and Home Economics Experiment Station.

Fulton, Murray, and Konstantinos Giannakas. 2001. "Agricultural Biotechnology and Industry Structure," *AgBioForum* 4, 2: 137-151.

Graff, Gregory D., Gordon C. Rausser, and Arthur A. Small. 2003. "Agricultural Biotechnology's Complementary Intellectual Assets," *Review of Economics and Statistics* 85, 2 (May, 2003): 349-363.

Heisey, Paul, and David Schimmelpfennig. 2006. "Regulation and the Structure of Biotechnology Industries," in Richard E. Just, Julian M. Alston, and David Zilberman (eds.), *Regulating Agricultural Biotechnology: Economics and Policy*, New York: Springer.

Heisey, Paul W., C.S. Srinivasan, and Colin Thirtle. 2001. *Public Sector Plant Breeding in a Privatizing World*, Agriculture Information Bulletin Number 772, U.S. Department of Agriculture, Economic Research Service, available at www.ers.usda.gov/ publications/aib772/

Howard, Philip H. 2009. "Visualizing Consolidation in the Global Seed Industry: 1996-2008," *Sustainability* 1(4): 1266-1287, available at: www. mdpi.com/journal/ sustainability

Huffman, Wallace E., and Robert E. Evenson. 2006. *Science for Agriculture: A Long-Term Perspective, 2nd edition*, Ames, IA: Iowa State University Press.

International Seed Federation. "Estimated Value of the Domestic Seed Market in Selected Countries, International Seed Federation, Seed Statistics," Nyon, Switzerland, Current estimates available at: www.worldseed.org/isf/seed

James, Clive, 2007. *Global Status of Commercialized Biotech/GM Crops: 2007*, ISAAA Brief No. 37, International Service for the Acquisition of Agri-Biotech Applications, Ithaca, NY.

Just, Richard E. 2006. "Anticompetitive Impacts of Laws That Regulate Commercial Use of Agricultural Biotechnologies in the United States," in Richard E. Just, Julian M. Alston, and David Zilberman (eds.), *Regulating Agricultural Biotechnology: Economics and Policy*, New York: Springer.

Just, Richard E., and Darrell L. Hueth. 1993. "Multimarket Exploitation: The Case of Biotechnology and Chemicals," *American Journal of Agricultural Economics* 75, 4 (November): 936-945.

Kalaitzandonakes, Nicholas, and Bruce Bjornson. 1997. "Vertical and Horizontal Coordination in the Agro-Biotechnology Industry: Evidence and Implications," *Journal of Agricultural and Applied Economics* 29(1): 129-139.

Kalaitzandonakes, Nicholas, Julian M. Alston, and Kent J. Bradford. 2006. "Compliance Costs for Regulatory Approval of New Biotech Crops," in Richard E. Just, Julian M. Alston, and David Zilberman (eds.), *Regulating Agricultural Biotechnology: Economics and Policy*, New York: Springer.

Kalaitzandonakes, Nicholas, Alexandre Magnier, and Douglas Miller. 2010. "A Worrisome Crop? Is There Market Power in the U.S. Seed Industry?" *Regulation* 33(4): 20-26.

King, John, and Paul Heisey. 2003. "Ag Biotech Patents: Who's Doing What?" *Amber Waves*, Vol. 1, Issue 5, U.S. Department of Agriculture, Economic Research Service, pp. 12-13, available at: www.ers.usda.gov/ amberwaves/november03/datafeature/

King, John, and David Schimmelpfennig. 2005. "Mergers, Acquisitions, and Stocks of Agricultural Biotechnology Intellectual Property," *AgBioForum* 8 (2&3): 83-88.

Le Buanec, Bernard. 2007. "Evolution of the Seed Industry During the Past Three Decades," *Seed Testing International* (ISTA News Bulletin) No. 134 (October): 6-10.

Magnier, Alexandre, Nicholas Kalaitzandonakes, and Douglas J. Miller. 2010. "Product Life Cycles and Innovation in the U.S. Seed Corn Industry," *International Food and Agribusiness Management Review* 13, 3: 17-36.

Marco, Alan C., and Gordon G. Rausser. 2008. "The Role of Patents Rights in Mergers: Consolidation in Plant Biotechnology," *American Journal of Agricultural Economics* 90, 1 (February 2008): 133-151.

McCouch, Susan R. 2001. "Is Biotechnology an Answer?" in K. Wiebe, N. Ballenger, and P. Pinstrup-Anderson (eds.), *Who Will be Fed in the 21st Century?* International Food Policy Research Institute, pp. 29-40.

Monsanto. 2009. "Observations on Competition in the U.S. Seed Industry," available at www.monsanto.com/newsviews/Pages/monsanto-submissiondoj.aspx (accessed October 2010).

Morris, Michael L., Joseph Rusike, and Melinda Smale. 1998. "Maize Seed Industries: A Conceptual Framework," in Michael L. Morris (ed.), *Maize Seed Industries in Developing Countries*, Boulder, CO: Lynne Rienner, pp. 35-56.

Moss, Diana L. 2009. *Transgenic Seed Platforms: Competition Between a Rock and a Hard Place?* American Antitrust Institute White Paper.

National Research Council (NRC). 2010. *The Impact of Genetically Engineered Crops on Farm Sustainability in the United States*, Washington, DC: National Academies Press.

Pray, Carl E., Jikun Huang, Ruifa Hu, Qihuai Wang, Bharat Ramaswami, and Prajakta Bengali. 2006. "Benefits and Costs of Biosafety Regulation in India and China," in Richard E. Just, Julian M. Alston, and David Zilberman (eds.), *Regulating Agricultural Biotechnology: Economics and Policy*, New York: Springer.

Schimmelpfennig, David E., Carl E. Pray, and Margaret F. Brennan. 2004. "The Impact of Seed Industry Concentration on Innovation: A Study of US Biotech Market Leaders," *Agricultural Economics* 30, 2: 157-167.

Shi, Guanming, Jean-Paul Chavas, and Kyle W. Stiegert. 2009. "Pricing of Herbicide-Tolerant Soybean Seeds: A Market Structure Approach," *AgBioForum* 12(3&4): 326-333.

Shi, Guanming, Jean-Paul Chavas, and Kyle Stiegert. 2010a. "An Analysis of the Pricing of Traits in the U.S. Corn Seed Market," *American Journal of Agricultural Economics* 92(5): 1324-1338.

Shi, Guanming, Kyle Stiegert, and Jean-Paul Chavas. 2010b. *An Analysis of Pricing in Horizontal and Vertical Markets: The Case of the U.S. Cottonseed Market*, FSRG Working Paper 2010-5, University of Wisconsin-Madison, Food System Research Group.

Shoemaker, Robbin. 2001. *Economic Issues in Agricultural Biotechnology*, Agricultural Information Bulletin No. 762, U.S. Department of Agriculture, Economic Research Service.

Stiegert, Kyle W., Guanming Shi, and Jean Paul Chavas. 2010. "Innovation, Integration, and the Biotechnology Revolution in U.S. Seed Markets," *Choices* 25, 2: 5-16.

Traxler, Greg. 1999. "Balancing Basic, Genetic Enhancement and Cultivar Development in an Evolving U.S. Plant Germplasm System," *AgBioForum* 2, 1: 43-47.

U.S. Department of Agriculture, Economic Research Service (USDA/ERS). Agricultural Biotechnology Intellectual Property Database, available at: www.ers.usda.gov/data/agbiotechip/

U.S. Department of Commerce, Technology Administration, Bureau of Industry and Security. 2003. *A Survey of the Use of Biotechnology in U.S. Industry*, October.

Virginia Polytechnic Institute and State University. Search USDA Releases and Notifications, Information Systems for Biotechnology, Blacksburg, VA, available at: www.isb.vt.edu/Default.aspx.

PRIVATE RESEARCH AND DEVELOPMENT FOR CROP PROTECTION CHEMICALS

John L. King and Keith O. Fuglie

Crop protection in agriculture improves yields by reducing crop losses from diseases and pests, especially insects, weeds, and fungi.[1] Application of synthetic chemical pesticides is a major crop protection activity, with agriculture accounting for 62 percent of pesticide sales in the United States (EPA, 2011) and 86 percent of pesticide sales worldwide (AGROW, 2007b).

Crop protection also includes the application of pesticides derived from natural sources ("biopesticides") and the use of integrated pest management (IPM) techniques, such as the introduction of pest predators, the use of traps baited with insect pheromones, and the use of physical pest barriers.

Chambers and Lichtenberg (1994) estimate that U.S. crop losses from pests decreased from 15 percent of crop value in the 1950s to about 3 percent by the 1980s due to increasing use of chemical pesticides. More recently, GM crops with herbicide tolerance and insect resistance confer a degree of inherent crop protection capability in varieties with those traits.

R&D spending by the crop protection industry worldwide has increased only slightly in nominal terms (unadjusted for inflation) over the past few decades and fell in real terms (inflation adjusted) over 1994-2010.

New agricultural chemical products resulting from R&D have reached the market during this period but at a slower rate of introduction than in previous years. At the same time, the crop protection industry has undergone a transformation with the widespread planting of GM crops. Some firms have taken steps to restructure their R&D programs to integrate biotechnology-based crop protection with chemical pest control.

Market Size and R&D Spending in the Global Pesticide Industry

Sales of crop protection chemicals ranged between $30 billion and $40 billion for most of the 1990s and 2000s. In nominal terms, global agricultural chemical sales increased from $32.7 billion in 1994 to $47.1 billion in 2010; the annual growth rate of 2.3 percent was slightly higher than the rate of inflation.[2] In addition to showing little real growth in overall market sales, global pesticide use during the same period declined relative to the growth of crop output, both in terms of physical volume of active ingredients applied and the value of product sales (fig. 3.1).

Research spending and sales revenue by the 45 largest global agricultural chemical companies in 2006 are shown in table 3.1. The table also breaks down sales and R&D spending by region according to where these companies are incorporated. A number of other firms that appear to conduct little if any R&D, most of them quite small and located in China and India, are not included in the table.[3]

In 2006, companies based in just four countries—Germany, the United States, Switzerland, and Japan—accounted for 83 percent of global sales of agricultural chemicals. These countries are home to some of the largest multinational agricultural chemical producers, with a significant share of their products manufactured and sold in other regions. For example, Germany, the location of the corporate headquarters of two of the largest companies (BASF and Bayer), accounted for 40 percent of total global R&D in agricultural chemicals. Eight U.S.-based companies accounted for 23 percent of the global total in 2006; together, they spent nearly $600 million. In addition to engaging in global sourcing and sales of agricultural chemicals, multinational agricultural chemical firms conduct R&D to develop and test new chemicals at a global network of research facilities. For example, the Swiss-based company Syngenta has chemical discovery and development laboratories in Switzerland, the UK, the United States, and India; biotechnology research centers in the United States and China; and a global network of crop breeding and field research stations (Syngenta, 2010).

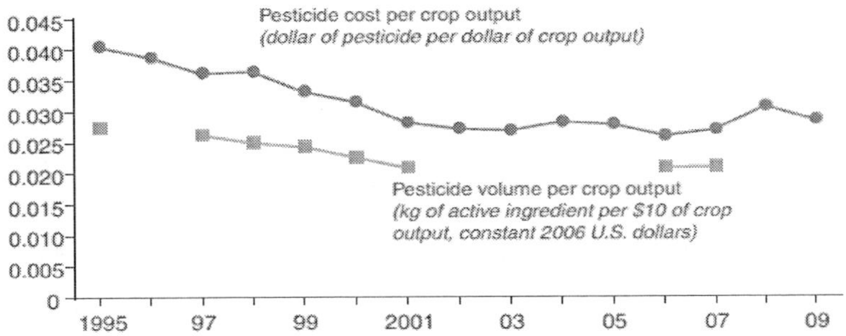

Sources: USDA, Economic Research Service. Pesticide sales for crop uses are from Cropnosis. Estimates of total active ingredients are from U.S. Environmental Protection Agency (1997, 1999, 2002, 2004, and 2011). The value of global crop production is from Food and Agriculture Organization of the United Nations, which reports the gross aggregate crop output measured in constant 2000 U.S. dollars. Value of crop output and pesticide use is converted to constant 2006 U.S. dollars using the U.S. Gross Domestic Product implicit price deflator (Economic Report of the President, 2009).

Figure 3.1. Pesticide use intensity in global agriculture.

Table 3.1. Crop protection chemical research and development (R&D) spending and sales by country in 2006

Region where company is incorporated	Companies	R&D	Sales	Average R&D intensity	Industry R&D share	Global market share
	Number	*Mil. U.S. dollars*		*Percent*		
North America	8	599	9,294	6.4	23	25
Europe-ME	9	1,596	19,016	8.4	61	51
Asia-Pacific	27	404	7,925	5.1	16	21
Latin America	1	13	877	3.9	1	2
Global total	45	2,611	37,112	7.1	100	100

Note: R&D and sales are totaled over 45 companies included in our survey.

Sources: USDA, Economic Research Service using firms' financial reports, AGROW Reports, and confidential information provided by selected firms.

Structure of R&D in the Global Crop Protection Industry

The crop protection industry engages in several different types of R&D activity, including new product discovery research, product development, and marketing of new products from existing chemical active ingredients. More recently, firms have also pursued research aimed at providing crop protection through a combination of using agricultural chemicals with seeds with GM traits. Not all firms engage in each type of research. Some firms engage in new product discovery, some focus on the manufacture of off-patent products or patented products under license, and some "integrated" firms invest in both biological and chemical technologies (including discovery research) for crop protection.

Integrated firms that conduct both chemical and genetic research for crop protection engage in the highest level of R&D in terms of both absolute expenditures and expenditures relative to sales (table 3.2). (Monsanto is included in this group in recognition of its large market share in crop protection genetic traits and development of the herbicide glyphosate, although Monsanto no longer conducts significant chemical R&D.) Table 3.2 combines sales and expenditures for seed, trait, and chemical R&D for these integrated firms because many of them no longer report these activities separately. Syngenta, an integrated firm, recorded the highest revenue from agricultural chemical and total agricultural input sales of any company worldwide in 2010. Sumitomo was a leader among agricultural chemicals firms that conduct discovery research. Nufarm and Makhteshim-Agan topped the list of high-revenue agricultural chemical firms that did not conduct significant product discovery research.

Proprietary Versus Generic Products

Manufacturers of crop protection chemicals distinguish themselves by whether or not they develop new, proprietary crop protection compounds. New product development can significantly affect the amount of R&D conducted by a firm. This chapter classifies companies as "discovery" companies if they undertook research in new chemical discovery and "generic" if they did not engage in such work. Discovery research includes the search for new active ingredients and technologies. In discovery research, firms experiment to identify chemicals with beneficial properties or lower costs of production. Beneficial properties of new products include greater effectiveness in a specific crop species, against a difficult plant disease or class of pests, or with a different mode of action that can be combined with other crop protection efforts. When discovery research is successful, a firm can apply for patents to obtain the exclusive rights to use, manufacture, or sell its new products and processes. Firms then embark on development, including demonstration of the efficacy, safety, and environmental behavior of pesticides that is necessary to secure and maintain regulatory approval. Development also includes providing different formulations of active ingredients to buyers (e.g., offering active ingredients in different concentrations or in combination with other chemicals).

Table 3.2. Leading firms in the global crop protection industry in 2009/10

Company	Country of incorporation	Total ag R&D spending[1]	Ag chemical sales	Total agricultural seed, trait & chemical sales	Ag R&D/ag sales
Integrated chemical-biological crop protection firms			Mil. U.S. dollars		Percent
Syngenta	Switzerland	1,032	8,878	11,641	8.9
Bayer CropScience	Germany	955	7,284	9,057	10.5
BASF	Germany	720	5,348	5,348	13.5
Dow AgroSciences	U.S.	n.a.	3,708	4,341	n.a.
Dupont	U.S.	874	2,453	9,084	9.6
Monsanto	U.S.	1,205	2,029	9,640	12.5
Leading chemical discovery firms					
Sumitomo	Japan	220	2,458		8.9
FMC	U.S.	81	1,242		6.5
Arysta	Japan	n.a.	1,276		n.a.
Leading agricultural chemical generics firms					
Nufarm	Australia	35	2,357		1.5
Makhteshim-Agan2	Israel	23	2,180		1.1
Cheminova	Denmark	n.a.	996		n.a.
United Phosphorus Ltd.	India	n.a.	390		n.a.
Sipcam Oxon	Italy	n.a.	409		n.a.
Albaugh	U.S.	n.a.	n.a.		n.a.
n.a. = not available.					

[1]Total agricultural research and development (R&D) expenditures include research on agricultural chemicals and for integrated crop protection firms, research on crop biotechnology, and seed.

[2]Makhteshim-Agan was acquired by the China National Chemical Corp. (ChemChina) in 2011.

Source: USDA, Economic Research Service using company annual financial reports. Data are for 2010 except for Dow AgroSciences and Sipcam Oxon, both 2009.

Because of regulatory and marketing requirements, all manufacturers of crop protection chemicals make development expenditures, even for chemicals without patent protection that are sold as branded or unbranded generic products. Research by generic firms focuses on the development and testing components of chemical R&D.

Many companies, including discovery, generic, and integrated firms, produce both proprietary and nonproprietary products. Discovery firms may continue to manufacture older products that have gone off-patent, and generic firms may manufacture patented products under license from the patent holders. Thus, the market distinction is not exact, and total sales figures of discovery and integrated firms may include generic chemicals. Discovery firms, however, differ from generic firms both qualitatively and quantitatively in that they conduct different types of research and perform them at a higher degree of intensity. This chapter finds that discovery firms typically spend 7-10 percent of their product sales on R&D while generic firms spend 1-3 percent.

The Emergence of "Integrated" Crop Protection Companies

As described in chapter 2 of this report, the emergence of "life science" companies that integrate chemical and biological sciences for agricultural applications was a major development in the global agricultural input industries. This helped lead to consolidation among agricultural seed and chemical companies. As firms conducted research to adapt biotechnology for agriculture, they embarked on a new wave of mergers and acquisitions. Firms in the chemical sector, many of which had existing pharmaceuticals businesses, combined through mergers and acquisitions in an attempt to create life science firms able to leverage advances in biotechnology across agriculture and human health. These combinations reinforced other long-term trends toward consolidation already underway in the chemical industry (discussed later in this section).

These life science arrangements, however, ultimately proved unwieldy (Pray et al., 2005), and pharmaceutical firms largely divested their agricultural chemicals operations. Some of the remaining firms maintained a focus on agriculture and found themselves with operations in agricultural chemicals, trait development, and seed distribution. The emergence of the Big 6 companies (discussed in chapter 2) created "crop protection" firms able to integrate new GM seed varieties with complementary chemical products. With large market shares in chemicals, seed varieties, and technology, and possible economies of scale in R&D, production, and marketing, these firms are particularly well positioned to provide chemical and seed inputs during a time of high production and prices for grain, cotton, and other agricultural commodities.

Figure 3.2 traces the evolution of two integrated crop protection companies, Syngenta and Dupont.[4] Dupont's agricultural R&D program consisted mostly of chemical technology prior to 1997, but it expanded to include biological technology with its acquisition of Pioneer Hi-Bred in 1999. At the time, Pioneer Hi-Bred was the world's largest seed company. Subsequently, Dupont gained further technology and market assets through acquisitions of Verdia, a biotechnology firm, and Griffen, a midsized agricultural chemical manufacturer.

Syngenta was formed in 2000 through the merger of two companies, AstraZeneca and Novartis, both of which had already built up substantial capacities in agricultural life sciences from previous mergers and acquisitions. Between 2004 and 2010, Syngenta bought several additional seed companies to expand its product portfolio in field crops, horticulture, and sugar beets.

Monsanto's history reflects the changing nature of crop protection and the agricultural chemicals business. Monsanto invented and patented the broad-spectrum herbicide glyphosate, which is widely used in agriculture, especially to control weeds prior to the emergence of crops in the field. With the commercial release of GM crops used in conjunction with glyphosate, Monsanto's sales of agricultural chemicals increased significantly. However, the company then shifted its R&D focus toward traits and germplasm and largely discontinued further research in agricultural chemicals. After patent protection for glyphosate expired, Monsanto continued production of this chemical but has otherwise largely exited crop protection chemicals.[5]

Other companies, including Bayer CropScience, Dow AgroSciences, and BASF, have followed the pattern of using mergers and acquisitions to expand their agricultural R&D to include both chemicals and biotechnology. BASF has charted a course different from that of its competitors in that it has not acquired seed companies. Rather, it has focused its R&D on chemical and biological trait discovery. In 2007, BASF and Monsanto agreed to a plan to jointly develop new crop technologies, with Monsanto marketing products from the collaboration and both companies sharing net profits.

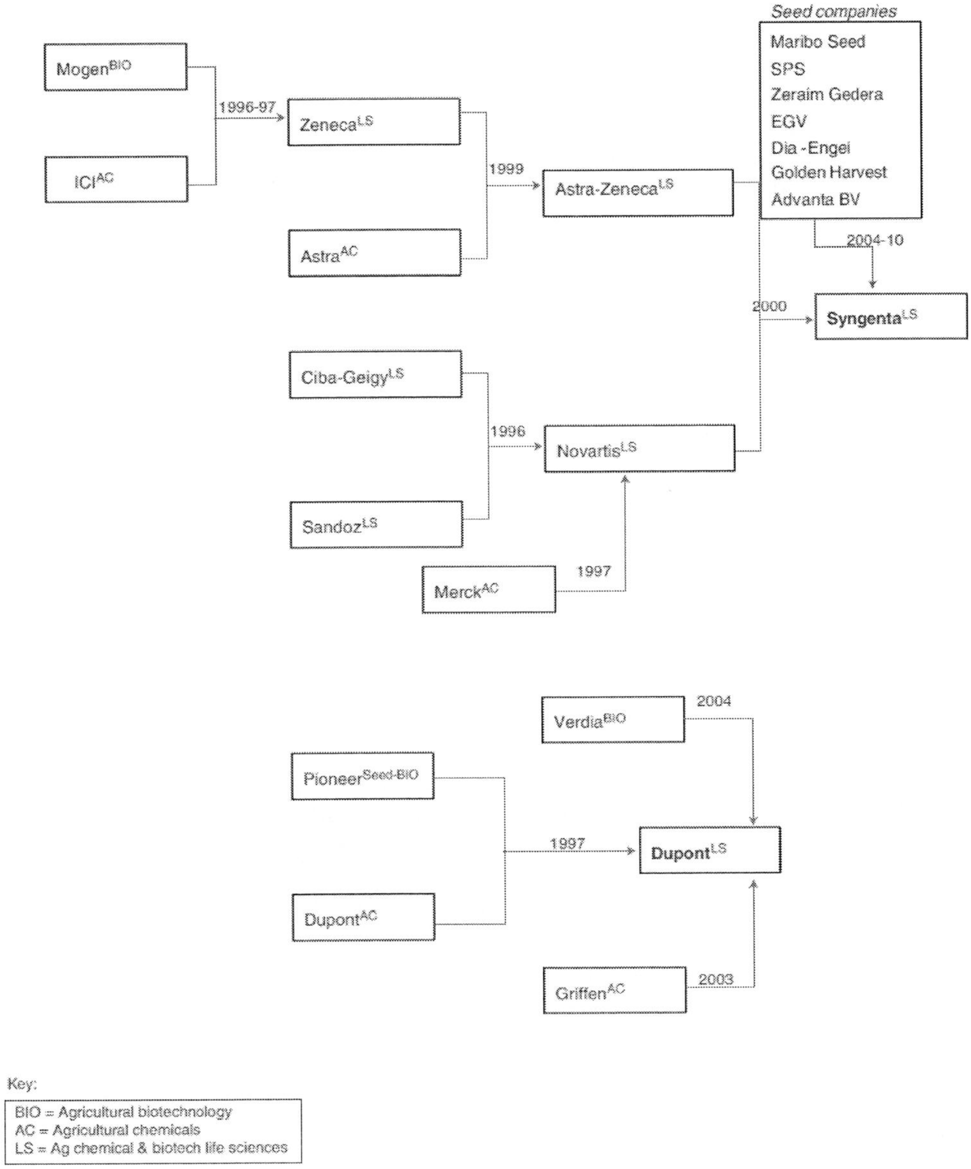

Source: USDA, Economic Research Service using data from Copping (2003), Fernandez-Cornejo (2004), and company websites. See also figure 2.2 for the formation of BASF, Bayer CropScience, Dow Chemical, and Monsanto.

Figure 3.2. Formation of agricultural life sciences companies: Syngenta and Dupont.

Table 3.3 decomposes industry R&D on agricultural chemicals and compares the research intensity of integrated, discovery, and generic firms. Between 1994 and 2010, annual spending on agricultural chemical R&D by the industry increased from $2.3 billion to $3.1 billion (1.9 percent per year), which was less than the rate of inflation. The life science companies account for about 75 percent of the sector's total agricultural chemical R&D, and both the life science companies and the discovery companies have much higher research intensities than the generic companies. Over time, however, R&D intensity for the life sciences and discovery firms declined slightly, each falling by about 1 percentage point over the period. Research intensity by generic firms has remained constant at about 1.6 percent of sales. This figure essentially reflects the costs of developing and registering product formulations and ongoing product maintenance.

Table 3.3. Research and development (R&D), sales, and concentration in the global crop protection industry

			Ag chemical R&D by type of firm[1]			Ag chemical R&D intensity by type of firm[1]		
Year	Global agricultural chemical market sales	Total industry agricultural chemical R&D	Integrated chemical-biotech	Other discovery	Generics*	Integrated Chemical-biotech	Other discovery	Generics*
	Million nominal U.S. dollars					*R&D/sales (%)*		
1994	32,735	2,268	1,612	612	44	8.7	8.0	1.8
1995	35,605	2,362	1,678	640	44	8.2	7.5	1.9
1996	36,765	2,494	1,782	664	48	8.2	7.9	1.8
1997	35,529	2,606	1,916	636	53	8.5	7.9	1.8
1998	36,476	2,606	1,975	574	57	8.7	7.4	1.9
1999	34,824	2,550	1,923	567	61	9.0	7.5	1.8
2000	34,353	2,321	1,671	588	62	8.2	8.3	1.9
2001	31,887	2,231	1,774	395	62	8.4	7.2	1.8
2002	31,248	2,044	1,613	370	62	8.9	6.9	1.7
2003	32,466	2,425	1,894	462	69	8.9	7.7	1.8
2004	36,758	2,594	2,030	490	74	8.6	7.4	1.6
2005	37,788	2,643	2,051	514	78	8.6	7.4	1.5
2006	37,112	2,596	1,997	508	91	8.5	7.3	1.9
2007	41,267	2,717	2,123	506	88	8.0	7.1	1.6
2008	50,023	2,975	2,294	570	111	7.0	6.9	1.6
2009	46,791	2,949	2,274	564	111	7.4	6.9	1.6
2010	47,144	3,077	2,348	617	113	7.8	7.1	1.6

[1]Integrated chemical-biotech firms conduct R&D in both agricultural chemicals and crop biotechnology. All of these (with the exception of Monsanto) engage in new chemical discovery. Other discovery firms also develop new kinds of chemical active ingredients but do not conduct significant seed-biotechnology research. Generic firms do not engage in new product discovery and manufacture off-patent products or products under license from patent holders. The R&D figures reported in the table only refer to the agricultural chemical R&D investments by these firms and exclude seed and biotechnology research.

Sources: USDA, Economic Research Service. Global agricultural chemical sales are from AGROW (2007) and include noncrop uses. Firm-level sales and R&D expenditures were compiled by ERS for 50 leading firms that manufactured agricultural chemicals between 1994 and 2007.

Sources of firm-level data include firm's financial reports, AGROW Reports, and confidential information provided by selected firms.

Rising Concentration in the Agricultural Chemical Industry

The supply of agricultural chemicals is fairly concentrated among several large producers and has become more so over time. Concentration ratios—the market share of the largest firms—have steadily increased over the past few decades. Between 1994 and 2009, the four-firm concentration rate (i.e., the market share of the largest four firms) rose from 28 percent to 53 percent of global pesticide sales (see table 1.7). Although AGROW (2007a) finds that at least 100 companies supply agricultural chemicals, by 2009, the 12 largest firms accounted for about 86 percent of the global market.

The rising concentration of the global crop production industry is the result of consolidation through mergers, acquisitions, and firm exits. This longterm trend toward consolidation is related to the lack of real sales growth. In nominal dollars, global sales of agricultural chemicals increased from $32.7 billion in 1994 to $47.4 billion in 2010, or an average of 2.3 percent per year, just slightly above the rate of inflation over the same period (table 3.3). Flat overall sales suggest that individual firm growth resulting from production efficiency, quality improvements, or new product introductions came at the expense of competitors.

High fixed costs associated with manufacturing also helped contribute to consolidation in the agricultural chemicals industry. Because chemical pesticide manufacturing is a capital-intensive industry, it favors fewer, larger firms that can spread fixed capital costs over a larger sales volume. Ollinger and Fernandez-Cornejo (1998) find evidence that high sunk fixed costs in the 1970s and 1980s led some firms to exit the industry. Of the 38 firms the study identified as having agricultural chemical R&D capacity during the period, 30 are now owned by the 6 largest discovery pesticide companies. Ollinger and Fernandez-Cornejo also find that larger firms and, especially, firms with higher international sales were more likely to expand and acquire other firms, which is consistent with the explanation that remaining firms derive a competitive advantage from spreading greater fixed costs over higher sales in more markets. In addition to being more likely to acquire other firms, large firms with higher international sales were less likely than other firms to be acquired.

Further support for the importance of fixed costs for consolidation in the crop protection chemicals industry comes from establishment-level data of manufacturing firms. Because capital and equipment used to manufacture pesticides are durable, capacity at existing manufacturing establishments has been able to meet declining volumes of production[6] without significant net capital formation, a deterrent to entry of new firms. The number of agricultural chemical manufacturing establishments in the United States has remained at approximately 250 for the past two decades (U.S. Department of Commerce, 2008), and the number across the EU was higher but stable at approximately 600 establishments (EUROSTAT). Manufacturing establishments in the EU and the United States, which account for most agricultural chemical production worldwide, have also employed fewer employees and earned lower revenues per establishment over time.

R&D for discovery of new active ingredients is another fixed cost that must be recovered over the life of a product. Firms often rely on intellectual property protection, such as patents, to recover these costs, which otherwise would be a disincentive for performing R&D. In the absence of patent protection for new products, generic producers can offer identically formulated products at lower prices because they are unburdened by sunk R&D costs. Patents pose a legal threat to imitators and enable firms that create successful new products to sell at higher prices and profit margins than they would earn without patent protection. Patented products can confer other first-mover advantages, such as recognition of trademarked brands that persist even after the expiration of patent protection. Producers of branded and unbranded generic crop protection chemicals still must comply with the regulatory and product safety requirements that are part of the fixed costs of pesticide registration. However, the United States, the EU, and other countries also grant intellectual property rights to data submitted for regulatory review of new pesticide products. Registrants of new pesticide products retain multiyear exclusive rights to product registration and safety data and, subsequently, can require compensation by other users of the data.

Innovation Trends in Crop Protection

R&D expenditures on agricultural chemicals have been shifting toward development expenses and away from discovery research. With less research on new chemical discovery, one might expect fewer new product innovations over time, and data from the U.S. Environmental Production Agency (EPA) and U.S. Patent Office appear to confirm this.

Regulations and the Composition of Agricultural Chemical R&D

In addition to facing greater technical requirements for successful discovery, firms registering new active ingredients face higher regulatory costs. Ollinger and Fernandez-Cornejo (1995) describe the increasing strictness of U.S. pesticide regulation in the 1970s and 1980s. They find that inflation-adjusted testing costs increased by an average of about 15 percent per year during this period, resulting in a decrease in new pesticide registrations of 6.75 percent per year (holding other factors constant). Regulations have also decreased the availability of older pesticide products. For example, re-registration requirements in the United States in 1988 and in the EU in 1991 removed especially hazardous chemicals from the market when less toxic substitutes were available.[7]

Due in part to these higher regulatory burdens on the use of agricultural chemicals, firms have targeted crop protection R&D toward such product qualities as increased effectiveness per application, reduced toxicity to humans and other nontarget species, and less persistence of harmful chemicals in the environment. Fernandez-Cornejo and Jans (1995) find that after adjusting for these improvements to quality, pesticide prices fell for U.S. corn, soy, sorghum, and cotton production between 1967 and 1992. Lower sales also reflect lower application rates for a given amount of pest protection. Ollinger and Fernandez-Cornejo (1995) estimate that increases in total regulatory testing costs (which rose 15 percent per year on average) were associated with 4.2 percent more registrations per year in a "less toxic" category even while new registrations were falling overall. And Paul et al. (2002) estimate that increasing use of higher quality pesticides in the United States is associated with reductions in human health risks from pesticide leaching and runoff. Although these studies suggest that the greater developmental and regulatory requirements have been somewhat successful at encouraging crop protection practices with fewer health and environmental risks, it is not clear that incentives and regulations encourage the optimum R&D investments for improving agricultural chemicals (Zilberman and Millock, 1997). For example, higher fixed costs of regulation might lead the industry to focus R&D on more complex or difficult compounds with greater commercial potential, or to abandon R&D for smaller markets, such as those involving specific fruit and vegetable crops (Ollinger and Fernandez-Cornejo, 1995).

Confirming the importance of growing developmental costs in the crop protection chemicals industry, survey data collected by the consulting firm PhillipsMcDougall (2010) reveal that regulatory compliance and testing now account for a majority of the full cost associated with introducing new active ingredients by the largest discovery-oriented firms. Development costs for new active ingredients registered at these firms were 57 percent of total R&D costs in 2005-08, up from 43 percent in 2000 (fig. 3.3). The timing of this change followed a period of intense industry consolidation (discussed in section 3.2) but was not associated with a simultaneous change in pesticide regulations. This sequence of events suggests that the shift of R&D costs toward development might reflect company strategies to research more complex and difficult targets. The cost of developing new active ingredients grew by more than 50 percent between 1995 and 2005-08 in real terms. This rate of change was only slightly slower than the rate of increase in new active ingredient development costs during the period of tightening regulatory requirements of the 1970s and 1980s (Ollinger and Fernandez-Cornejo, 1995).

New Product Registrations

The shift away from discovery expenses may stem in part from the growing difficulty of finding novel active ingredients that address unmet crop protection needs at a competitive price. Although crop pests constantly adapt to their environment and can eventually develop resistance to crop protection strategies, the cumulative effort of decades of crop protection R&D has produced effective, inexpensive solutions. Hartnell (1996) describes a "golden age" of agricultural chemical discovery in the mid-20th century characterized by rapid introduction of new active ingredients that overlapped with the reduction in crop losses from pests in the United States to as low as 3 percent per year (Chambers and Lichtenberg, 1994). The number of new active ingredients introduced in EPA pesticide registrations in the United States peaked at about 40 per year in the 1960s and has subsequently fallen to less than 10 per year every year since 1988 (fig. 3.4). The rate of introduction declined after cumulative research successes exhausted prominent commercial and techno logical opportunities, both in terms of new active ingredients and remaining uncontrolled pests. Firms began to screen increasingly larger numbers of candidate compounds to find a new chemical with marketable advantages for yield improvement or reduced toxicity. More recently, firms have employed new approaches to chemical discovery, including combinatorial chemistry, high-throughput screening, and computer simulation of molecular interaction (Hartnell, 1996; Joly and Lemarié, 2002). With these improved technologies, firms typically screen an average of 140,000 chemical compounds per new registered active ingredient, up from 52,500 in 1995 (PhillipsMcDougall, 2010). However, the observed slowdown in active ingredient introductions shown in figure 3.4 suggests that gains in efficiency have not kept pace with the increasing difficulty of identifying novel active ingredients that meet the rising bar for commercial introduction.

Pesticide Patents

The number of patent awards for pesticides provides another view of research and technology growth in crop protection chemicals. Figure 3.5 shows an estimate of the number of pesticide patents issued with an international patent classification of A01N.[8] This number has generally increased over the past three decades, although the approach of counting A01N patents might not reflect changes in the complexity of inventions or the value of their patent rights. Growth of patent awards has been more rapid in other patent classification areas, and the share of pesticide patents relative to all patent awards in the United States has been declining.

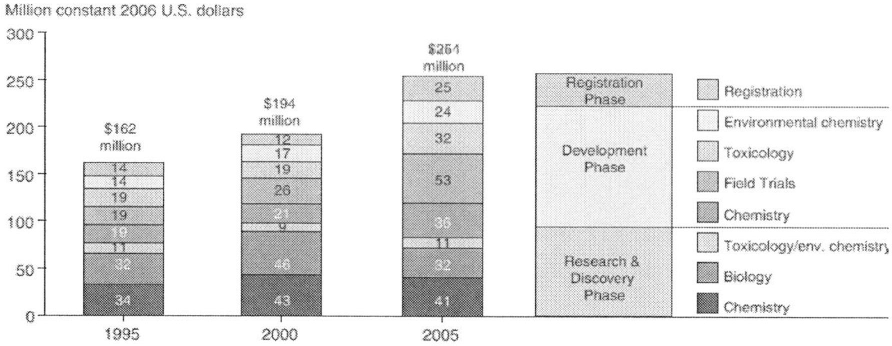

Source: USDA, Economic Research Service using data from PhillipsMcDougall (2010).

Figure 3.3. Costs of bringing a new agricultural chemical to the market.

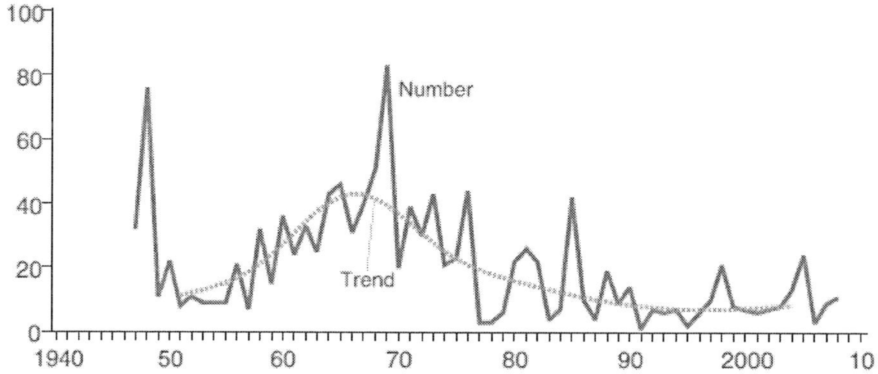

Note: Certain years exhibit spikes that are attributable to regulatory changes: the 1947 introduction of the registration requirement under the Federal Insecticide, Fungicide, and Rodenticide Act (FIFRA), the 1978 adoption of the Rebuttable Presumption Against Registration policy, and 1988 amendments to FIFRA that required re-registration of active ingredients in use in the United States.

Source: USDA, Economic Research Service analysis of data from the EPA Pesticide Product Information System.

Figure 3.4. New pesticide active ingredient registrations in the United States.

Adoption of Genetically Modifi Ed Crops

Over the first 15 years of commercial cultivation of GM varieties beginning in 1996, the principal applications of biotechnology in agriculture have been herbicide tolerance and insect resistance. Herbicide tolerance enables growers to spray fields with herbicides (especially glyphosate and glufosinate) to reduce weed pressure even after crops have emerged. Insect resistance is made possible through the introduction of genes that cause plants to secrete proteins that are harmful to certain kinds of crop pests. These "plant-incorporated pesticides" have been adapted to secrete a wider variety of proteins in different locations on plants (e.g., leaves and roots).

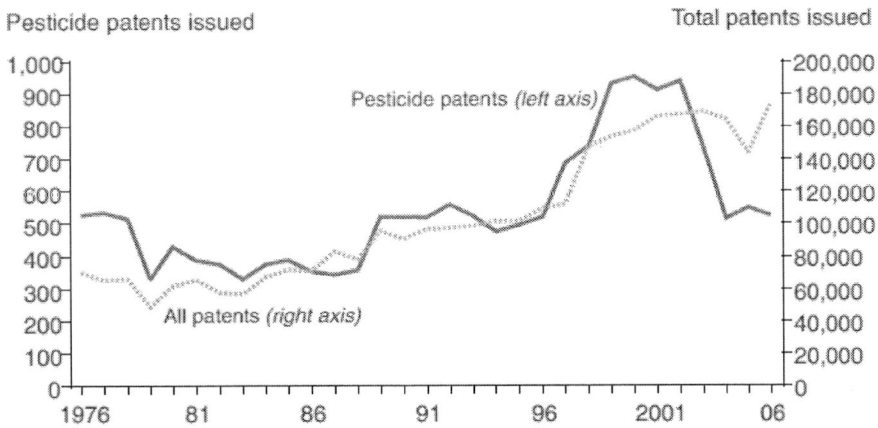

Source: USDA, Economic Research Service analysis of NBER Patent Data Project. A01N patents include "Preservation of bodies of humans or animals or plants or parts thereof; biocides (e.g., as disinfectants, as pesticides, or as herbicides); pest repellants or attractants; and plant growth regulators."

Figure 3.5. Trends in pesticide patents and total patents issued annually in the United States.

Both of these applications of biotechnology affect the level and makeup of demand for chemical crop protection products. Herbicide tolerance increases the demand for glyphosate and glufosinate but potentially reduces the need for other herbicides. Genetic insect resistance allows growers to control insect pests with fewer applications of synthetic chemicals. Between 1995 (just prior to the introduction of GM crops) and 2006, herbicides increased their share of the global crop protection market from 43 to 48 percent; over the same period, the insecticide share of the global crop protection market fell from 33 to 25 percent.[9]

Recent empirical studies examine the net effects of the adoption of GM crops on the quantities and patterns of agricultural chemicals. Benbrook (2004) finds that GM corn, cotton, and soybeans in the United States use more herbicides relative to conventional varieties of these crops, and that this increase in use outweighs the decrease in insecticide use in corn and cotton. Benbrook also notes that the development of herbicide resistance in weeds is likely to increase with more widespread use of herbicide tolerant varieties, a concern examined in more detail in a recent report by the National Research Council (2010). Fernandez-Cornejo and Caswell (2006) find evidence of net reductions in pesticide use and emphasize reported lower pesticide costs and easier pest management with GM varieties. Also, Fernandez-Cornejo and McBride (2002) emphasize that the pesticides used in conjunction with GM crops are less toxic and less environmentally persistent than alternative applications. Huang et al. (2002) find much lower use of pesticides in China in fields planted with insect-resistant GM cotton. Globally, Brookes and Barfoot (2008) find countries adopting GM varieties have lower pesticide use by weight of active ingredient than nonadopting countries, reducing their environmental impact. Another potential benefit of herbicide tolerance is a decrease in the need for weed control through mechanical tillage, which can cause soil erosion and runoff. However, Fuglie (1999) finds no significant difference in herbicide usage between fields with conservation tillage and those with conventional tillage in the U.S. Corn Belt.

Innovation in Biopesticides

A largely separate market development in the crop protection chemical industry is the growing use of biopesticides. Biopesticides are naturally occurring pest control agents. Their role in organic production gives them strong potential for growth in the marketplace. The Biopesticide Industry Alliance (2011) estimates that global biopesticide sales ranged as high as $1 billion in 2010, with a majority of sales attributed to orchard crops. Organic production standards do not permit the application of many synthetic chemical pesticides or GM crops, so organic growers often rely on biopesticides for crop protection. Biopesticide sales can be expected to increase because of growth in organic agriculture. U.S. organic cropland acreage reached 0.72 percent of total crop acreage in 2008 following average annual growth rates over 11 percent between 1992 and 2008 (USDA/ERS, 2007).

U.S. registrations of new biopesticide products vary by year of introduction (fig. 3.6).[10] Approximately 20 percent of registered biopesticide products are related to the bacterium Bacillus thurengiensis (Bt), many of which are for living Bt strains commonly formulated for pest control as a liquid or spray. For the past two decades, however, all new biopesticide registrations that use Bt have been GM crops that have the Bt gene that cannot be used in organic production.

Conclusion

Research and development in crop protection chemicals reflects several trends in the industry that have persisted for decades. Industry consolidation has continued over time as leading firms have engaged in mergers and acquisitions to achieve efficiencies in production, even as the number of firms producing agricultural chemicals remained roughly constant. In real terms, sales of agricultural chemicals remained somewhat constant over most of the 1990s and 2000s, with lower physical volume and intensity of use relative to the growing value of world agricultural production. New active ingredients in pesticide registrations have been slower to reach the market, in part because of increasing technical barriers, regulatory requirements, and lower incentives to develop and adopt chemicals that have off-farm health, safety, and environmental benefits. Some new products introduced during this period have been successful, but the production of branded and unbranded generic products with lower prices and profit margins also grew. Biopesticide sales have increased with the growth of organic agriculture but still account for less than 3 percent of the overall market.

Despite presenting a picture of a mature, consolidating industry that slow sales growth and consolidation would seem to imply, the crop protection chemicals industry has also undergone significant transformation due to the commercial introduction of GM varieties. Crops with engineered traits that confer insect resistance and herbicide tolerance affect the demand for agricultural chemicals, reducing demand for some insecticides and shifting demand toward the herbicides to which those crops are resistant. A few large multinational firms—which engaged in a large number of mergers, acquisitions, and company restructuring—have both capabilities in agricultural chemical manufacturing and the technology necessary to develop elite germplasm with crop protection traits. These firms are able to integrate both chemical and biological approaches to offer complementary types of crop protection. They compete with a number of large firms that maintain a traditional approach to chemical discovery and development and with firms that specialize in off-patent, generic products.

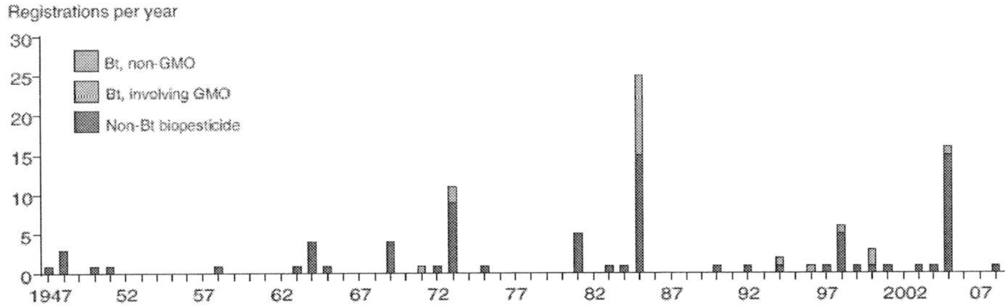

GMO = genetically modified organism.

Source: USDA, Economic Research Service analysis of data from EPA Pesticide Product Information System.

Figure 3.6. New registrations of biopesticide products in the United States.

References

AGROW Reports. 2001. *The Global Crop Protection Industry in 2010*, PJB Publications, Ltd., UK.

AGROW Reports. 2007a. *Agrow's Complete Guide to Genetic Pesticides: Volume I – The Companies (2007 Edition)*, London: Informa UK Ltd.

AGROW Reports. 2007b. *Agrow's Top 20: 2007 Edition*, Company Report Series, London: Informa UK Ltd.

Benbrook, C.M. 2004. "Genetically Engineered Crops and Pesticide Use in the United States: The First Nine Years," *BioTech InfoNet Technical Paper 7.*

Biopesticide Industry Alliance. 2011. "Biopesticide Facts: Market Overview." Available at www.biopesticideindustryalliance.org/marketinfo.php (accessed April 28, 2011).

Brookes, Graham, and Peter Barfoot. 2008. "Global Impact of Biotech Crops: Socio-Economic and Environmental Effects," 1996-2006, *AgBioForum* 11 (1): 21-38.

Chambers, Richard G., and Eric Lichtenberg. 1994. "Simple Econometrics of Pesticide Productivity," *American Journal of Agricultural Economics* 76 (3): 407-417.

Copping, Len. 2003. "The Evolution of Crop Protection Companies," *Pesticide Outlook* 14 (December): 276-78.

Economic Report of the President. 2009. U.S. Government Printing Office.

EUROSTAT. Statistics on the Production of Manufactured Goods. Manufacture of Pesticides and Other Agro-Chemical Products. European Commission, Luxembourg, available at: http://epp.eurostat.ec.europa.eu/statistics_explained/index.php/User:EXT-P-Rentze popoulos/Sandbox/Pesticide,_paint,_soap_and_fibre_production_statistics# Manufacture_of_pesticides_and_other_agro-chemical_products (accessed September 7, 2010).

Food and Agriculture Organization of the United Nations (FAO). FAOSTAT database, Rome, Italy, available at: http://faostat.fao.org/ (accessed October, 2010).

Fernandez-Cornejo, Jorge. 2004. *The Seed Industry in U.S. Agriculture*, Agricultural Information Bulletin No. 786, U.S. Department of Agriculture, Economic Research Service, available at: www.ers.usda.gov/ publications/aib786/

Fernandez-Cornejo, Jorge and Margriet Caswell. 2006. *The First Decade of Genetically Engineered Crops in the United States*, Economic Information Bulletin 11, U.S. Department of Agriculture, Economic Research Service, available at: www.ers.usda.gov/publications/eib11/

Fernandez-Cornejo, Jorge, and Sharon Jans. 1995. "Quality-Adjusted Price and Quantity Indices for Pesticides," *American Journal of Agricultural Economics* 77 (August): 645-659.

Fernandez-Cornejo, Jorge and William McBride. 2002. *Adoption of Bioengineered Crops.* Agricultural Economic Report 810, U.S. Department of Agriculture, Economic Research Service, available at: www.ers.usda.gov/publications/aer810/

Fuglie, Keith O. 1999. "Conservation Tillage and Pesticide Use in the Cornbelt," *Journal of Agricultural and Applied Economics* 31(01): 133-147.

Hartnell, Gaynor. 1996. "The Innovation of Agrochemicals: Regulation and Patent Protection," *Research Policy* 25(3): 379-395.

Huang, Jikun, Scott Rozelle, Carl Pray, and Qinfang Wang. 2002. "Plant Biotechnology in China," *Science* 295 (5555): 674-676.

James, Clive (2010). *Global Status of Commercialized Biotech/GM Crops: 2010*, ISAAA Brief Number 42, International Service for the Acquisition of Agri-Biotech Applications.

Joly, Pierre-Benoit, and Stéphane Lemarié. 2002. "The Technological Trajectories of the Agrochemical Industry: Change and Continuity," *Science and Public Policy* 29(4): 259-266.

National Research Council. 2010. *The Impact of Genetically Engineered Crops on Farm Sustainability in the United States*, Washington, DC: The National Academies Press.

Ollinger, Michael, and Jorge Fernandez-Cornejo. 1995. *Regulation, Innovation, and Market Structure in the U.S. Pesticide Industry*, Agricultural Economics Report No. 719, U.S. Department of Agriculture, Economic Research Service.

Ollinger, Michael and Jorge Fernandez-Cornejo. 1998. "Sunk Costs and Regulation in the U.S. Pesticide Industry," *International Journal of Industrial Organization* 16(2): 139-168.

Paul, C.J.M., V.E. Ball, R. Felthoven, A. Grube, and R.F. Nehring. 2002. "Effective Costs and Chemical Use in United States Agricultural Production: Using the Environment as a "Free" Input," *American Journal of Agricultural Economics* 84(4): 902-15.

PhillipsMcDougall. 2010. *The Cost of New Agrochemical Product Discovery, Development and Registration in 1995, 2000 and 2005-8*, Consultancy Study for Crop Life America and the European Crop Protection Association, available at: www.croplife.org/public/ reports.

Pray, C., J.F. Oehmke, and A. Naseem. (2005). "Innovation and Dynamic Efficiency in Plant Biotechnology: An Introduction to the Researchable Issues," *AgBioForum* 8(2&3), 52-63, available at: www.agbioforum.org.

Syngenta. Company website, available at: www2.syngenta.com/en/products_ brands/ research_where.html (accessed September 7, 2010).

U.S. Department of Commerce, Bureau of the Census (DOC). 2008. Annual Survey of Manufactures: General Statistics: Statistics for Industry Groups and Industries, NAICS325320.

U.S. Department of Agriculture, Economic Research Service (USDA/ERS). 2007. "U.S. Certified Organic Farmland Acreage, Livestock Numbers, and Farm Operations, 1992-2008," available at www.ers.usda.gov/data/ organic/data/farmlandlivestockandfarm.xls.

U.S. Environmental Protection Agency (EPA). Pesticide Product Information System, available at: www.epa.gov/pesticides/PPISdata/index.html (accessed July, 2009).

U.S. Environmental Protection Agency (EPA). 1997. "Pesticide Industry Sales and Usage: 1994 and 1995 Market Estimates," available at: www.epa.gov/ oppbead1/pestsales/

U.S. Environmental Protection Agency (EPA). 1999. "Pesticide Industry Sales and Usage: 1996 and 1997 Market Estimates." available at: www. epa.gov/oppbead1/pestsales/

U.S. Environmental Protection Agency (EPA). 2002. "Pesticide Industry Sales and Usage: 1998 and 1999 Market Estimates," available at: www. epa.gov/oppbead1/pestsales/

U.S. Environmental Protection Agency (EPA). 2004. "Pesticide Industry Sales and Usage: 2000 and 2001 Market Estimates," available at www. epa.gov/oppbead1/pestsales/

U.S. Environmental Protection Agency (EPA). 2011. "Pesticide Industry Sales and Usage: 2006 and 2007 Market Estimates," available at www.epa.gov/ oppbead1/pestsales/

Zilberman, D., and K. Millock. 1997. "Financial Incentives and Pesticide Use," Food Policy 22 (2): 133-144.

PRIVATE RESEARCH AND DEVELOPMENT FOR SYNTHETIC FERTILIZERS

David Schimmelpfennig, Keith Fuglie and Paul Heisey

Inorganic, synthetic fertilizer is a critical ingredient in the global food economy. In 2008, global consumption of the three main agricultural fertilizer nutrients—nitrogen (N), phosphate (P2O5), and potash (KCl, or potassium)—totaled 162 million metric tons (FAO).[1] Nitrogen accounts for about 63 percent of the total tonnage of fertilizers applied, phosphate another 21 percent, and potassium 15 percent.

At world trading prices for major fertilizers, the size of the global fertilizer market was about $68 billion in 2005 but soared to over $200 billion by 2008 due to a rapid rise in fertilizer prices.[2] Demand for fertilizers has risen in recent years, mostly in developing countries.

In high-income countries, where application rates are generally higher, fertilizer use has been stable or declining. In agricultural areas with high fertilization rates, environmental concerns stem from fertilizer runoff and leaching, which can affect surface and groundwater quality (USDA/ERS, 2006).

Nitrogen fertilizer can also vaporize into the atmosphere in the form of nitrous oxide (N_2O), which has been identified as a greenhouse gas and a contributor to global climate change (IPCC, 2007).

The Global Agricultural Fertilizer Market and Industry Structure

The fertilizer industry has undergone significant changes over the past half century. Park (2001) provides a comprehensive overview of the evolution of the global fertilizer industry during the 20th century and the structure of the industry as it stood in the late 1990s. Following World War II, production of chemical fertilizers increased rapidly, partly due to the conversion of munitions factories to fertilizer production. Many countries viewed fertilizer as a strategic industry, which led to significant government intervention in fertilizer markets, both in terms of direct ownership of factories and control of trade and prices. Since the 1980s, many fertilizer markets have been liberalized or privatized, although some governments continue to maintain a controlling interest in the industry. In the 1990s, fertilizer manufacturing and use in the countries of the former Soviet Union declined sharply, as did industry consolidation and company mergers in Europe and North America. By 2008, government-owned and government-controlled production accounted for 57 percent of global nitrogen fertilizer, 47 percent of phosphate fertilizer, and 19 percent of potassium fertilizer (PotashCorp, 2008).

The market structure that emerged from this period of liberalization and consolidation is markedly different for the three primary nutrients. The market structure for N fertilizers is the least concentrated globally. Manufacture and pricing of N is closely associated with availability and cost of natural gas, which is a main ingredient used to synthesize ammonia (the main feedstock for N fertilizer). Due to the high cost of transporting ammonia, most N fertilizer is consumed in or close to its country of manufacture, and more than 60 countries have manufacturing facilities for N fertilizers (PotashCorp, 2008). The United States is a net importer of ammonia, primarily from Trinidad, which has low-cost sources of natural gas.

The production of phosphate and potassium fertilizers is more concentrated due to the limited geographic availability of raw materials—phosphate rock and potash. These fertilizers are mined primarily from underground deposits. Phosphate rock is mined by both surface and underground methods, but surface mining is the predominant method used to mine phosphate deposits. Most potash mines are deep shaft mines, although a small share of the world's production also comes from salt lakes and seas. While about 40 countries produce phosphate fertilizers, just 5 account for 80 percent of global phosphate rock production (PotashCorp, 2008). Phosphate fertilizer manufacturing requires significant amounts of sulfuric acid, and production costs are sensitive not only to global prices of phosphate rock but also to prices of sulfur. Farm demand for phosphate fertilizer also faces competition from animal feed and industrial uses of phosphates.

Mineral resources of potash (used to produce potassium fertilizer) are even more concentrated than phosphate, with only three countries (Canada, Belarus, and Russia) accounting for 80 percent of the world's reserves (PotashCorp, 2008). Markets for phosphate and potassium fertilizers are more integrated globally than the market for N fertilizer. About 80 percent of global potash production is traded across international borders. World prices for fertilizers were fairly stable over 1995-2007 but rose significantly in 2007-08 (fig. 4.1). Factors contributing to the spike in fertilizer prices include a significant increase in world nutrient demand (as farmers responded to rising crop prices in this period), a sharp rise in the cost of energy and materials used in fertilizer manufacture (especially natural gas, sulfur, and phosphate rock), increased transportation costs, and the falling value of the U.S. dollar (Huang et al., 2009).

The Canadian company PotashCorp is the world's largest fertilizer manufacturer and produces significant amounts of all three primary macronutrients (table 4.1). In 2007, PotashCorp alone had about 22 percent of the total world production capacity in potash fertilizers. Moreover, PotashCorp, together with another Canadian firm Agrium and the U.S. firm Mosaic[3] (which sources most of its potash from mines in Canada), conducts its offshore marketing of potash through a common trading company Canpotex. This trading consortium controls more than one-third of global potash production. Another trading consortium, the Belarus Potash Company, handles the exports of Uralkali and Belaruskali, two major potash manufacturers in Eastern Europe. This small number of producers has historically been an important feature of the global fertilizer export market (Park, 2001) and enables it to exercise considerable market power, particularly in the potash fertilizer market. Even though the market structure for N fertilizers is the least concentrated globally, the market for urea fertilizer, an important type of N fertilizer, is more concentrated and this is reflected in regional price (Hernandez and Torero, 2011).

Eight firms account for just over half of global production capacity in phosphate fertilizers. The government-owned Moroccan company Office Cherifien des Phosphates (OCP) is by far the largest global producer of phosphates and is the source of about half of global exports of phosphate rock (PotashCorp, 2008). Production of ammonia, the feedstock for nitrogen fertilizers, is the least concentrated globally. The top seven producers account for only about 15 percent of global production. Looking forward, the global nitrogen industry is expected to remain relatively unconcentrated because recent technological advances in extracting natural gas from shale rock have caused estimates of economically recoverable gas reserves to increase and become more geographically diverse (Nature, 2009).

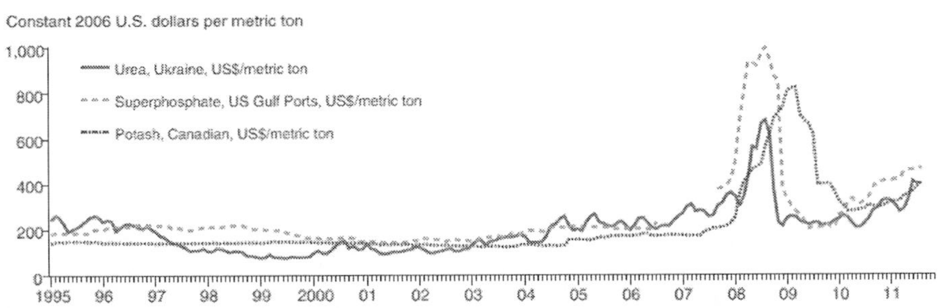

Source: USDA, Economic Research Service using monthly fertilizer prices from Haver Analytics and adjusted for inflation by the monthly U.S. Producer Price Index for Finished Goods, seasonally adjusted (Federal Reserve Bank of St. Louis).

Figure 4.1. World fertilizer prices.

R&D and Technical Change in the Fertilizer Industry

Over the course of the 20th century, a series of technical innovations led to a steady lowering of real fertilizer prices (Tomich et al., 1995). During 1909- 13, German chemists Fritz Haber and Carl Bosch developed the Haber-Bosch process to synthesize ammonia from air and a carbon feedstock and then convert ammonia to N fertilizer. In 1963, a centrifugal compressor replaced the complex reciprocating compressor, reducing the capital costs of ammonia synthesis by half (Tomich et al., 1995). New fertilizer formulations have also been developed that have increased nutrient density, such as urea (44-46 percent N) and triple superphosphate (45 percent P_2O_5), which effectively lowered the farm cost of fertilizer applications.

The sources of these innovations include university and government research laboratories and the research departments of private firms within and outside the fertilizer industry. A review of research spending by the fertilizer industry, however, reveals that these firms spend relatively little on R&D relative to company sales. Data compiled in a 1975 survey of private-sector agricultural R&D in the United States show that fertilizer manufacturers spent an average of only 0.21 percent of net sales on R&D (Wilcke and Williamson, 1977). Of the 42 fertilizer manufacturers in the dataset, only one, Yara International,[4] reported any R&D expenditures in its annual financial statement (Yara International, 2007). The R&D-to-sales ratio for Yara in 2006 was 0.25 percent, similar to the finding of the 1975 U.S. survey. It appears that most of the innovations in fertilizer manufacture are spillins from either the public sector or private firms in other chemical and energy industries or a result of "learning-by-doing" within the fertilizer industry.

Table 4.1. Fertilizer production capacities of largest companies in 2007
Capacity in million tons of primary product

Ammonia (NH_3)		Phosphate (P_2O_5)		Potash (KCl)	
Company	Capacity	Company	Capacity	Company	Capacity
Yara (Norway)	6.0	OCP (Morocco)	7.0	PotashCorp (Canada)	13.2
Terra Industries (U.S.)	4.5	Mosaic (U.S.)	4.6	Belaruskali (Belarus)	8.5
PotashCorp (Canada)	3.9	Agrium (Canada)	3.0	Mosaic (U.S.)	8.1
Agrium (Canada)	3.0	PotashCorp (Canada)	2.4	ICL (Israel)	6.0
CFI (U.S)	3.0	CFI (U.S.)	2.0	Silvinit (Russia)	5.4
IFFCO (India)	2.7	IFFCO (India)	1.7	Uralkali (Russia)	4.5
Mosaic (U.S.)	0.5	ICL (Israel)	1.0	Kali & Salz (Germany)	4.0
		GCT-CPG (Tunisia)	1.0	Sinofert (China)	3.0
				APC (Jordan)	2.0
				Agrium (Canada)	1.7
Capacity of listed companies	23.6		22.7		56.4
Total global capacity	154.3		43.0		67.0
Listed companies share of total	15%		53%		84%
Government-owned or subsidy-controlled production	47%		57%		19%

Source: USDA, Economic Research Service using PotashCorp (2007) and Heffer and Prud'homme (2008).

Several factors may account for the low R&D spending by the fertilizer industry. First, fertilizer is a large-volume and low-value commodity with few opportunities to develop differentiated products. The industry is capital intensive with major costs tied up in Greenfield development or raw material procurement. For example, the development of 2 million tons in new potash capacity is estimated to take 5-7 years and cost $2.8 billion (PotashCorp, 2008). For ammonia manufacture, up to 90 percent of the production cost is for natural gas. Under this cost structure, opportunities to reduce costs by developing more efficient manufacturing processes are limited. Second, the industry may lack incentives to develop more efficient fertilizers or fertilizer application methods (i.e., with less environmental escape). It may be difficult to claim intellectual property over this type of technology and therefore recoup returns to research investment. Further, the oligopoly structure of the fertilizer industry may reduce the competitive pressure on firms to innovate. More efficient fertilizers that capture a greater share of applied nutrients for plant growth could result in increased crop yields and agricultural production without a corresponding increase in nutrient use or even reduced farm demand (and industry revenue) for fertilizers. Such improvements in fertilizer formulations and application methods could have significant economic benefits to farmers as well as provide environmental benefits.

The fertilizer industry supports research on improving fertilizer use by jointly funding the International Plant Nutrition Institute (IPNI). The IPNI is a nonprofit, science-based organization that supports research and agronomic education about fertilizer utilization. It encourages adoption of best management practices to raise farm productivity as well as address environmental concerns associated with fertilizer use (IPNI, 2009). However, new innovations to improve agricultural nutrient management, such as soil testing and precision agriculture, have thus far come mainly from public and private sources outside of the fertilizer industry. The International Fertilizer Development Center (IFDC) is a public nonprofit R&D center that focuses on developing and transferring fertilizer technology to developing countries. The IFDC was established in 1974 as an outgrowth of the Tennessee Valley Authority's National Fertilizer Development Center (NFDC). The IFDC and its predecessor, NFDC, developed the majority of fertilizer products currently in use (International Fertilizer Development Center, 2010). In 2010, the IFDC launched a "Virtual Fertilizer Research Center," a global initiative to link researchers across universities and research laboratories to create a new generation of more efficient fertilizers and soil fertilizer management technologies. The IFDC is financially supported primarily by the (U.S. and foreign) public sector.

Given the lack of data on R&D spending by fertilizer companies, we estimate R&D spending by the industry as simply a fraction of sales. For fertilizers manufactured by firms in high-income countries, we assume an R&D-to-sales ratio of 0.25 percent. This is close to the average ratio for the U.S. fertilizer industry reported by Wilcke and Williamson (1977) and that reported by Yara International in 2006. For developing countries, we assume the fertilizer industry's R&D-to-sales ratio is half this rate, or 0.125 percent. Evenson and Westphal (1995, table 37.1, p 2242-3) find that average R&D intensities of industries in developing countries are half or less the average level in high-income countries. Production quantities of synthetic fertilizer nutrients (nitrogen, phosphate, and potassium) by country are from FAO. Value of production is estimated by multiplying production quantities by representative global fertilizer prices for urea, superphosphate, and potash, adjusted for nutrient content, as reported in the commodity price database of Haver Analytics. To estimate R&D in 2006-08, we use average fertilizer prices from 2002-05 instead of the inflated actual market prices.[5] Firms, particularly in an industry that does not conduct much research, are unlikely to change their R&D investment behavior in response to short-term price fluctuations. With these assumptions, we derive estimates of fertilizer R&D for each manufacturing country.

Among all countries, China has by far the largest fertilizer industry in the world and accounted for about one-fifth ($22.5 billion) of total global R&D in 2006 (table 4.2). North America (the United States and Canada) accounted for about 28 percent of global fertilizer

R&D. In several countries, governments still play a controlling role in domestic fertilizer markets, either by maintaining direct ownership stakes in fertilizer companies or controlling pricing, distribution, and trade in fertilizers. Among the five largest fertilizer-producing countries, government intervention predominates in two, China and India. In China, however, privately held share ownership in fertilizer companies is growing. Our estimate of R&D spending by the U.S. fertilizer industry, $19.1 million in 2006, is within the range of estimates provided by previous studies (table 4.3). Three studies conducted in the late 1970s estimated that fertilizer industry R&D in the United States was $6 million to $10 million annually, while a 1984 survey estimated the amount for that year at $35.3 million (constant U.S. 2006 dollars). While the considerable range of these estimates suggests uncertainty in the actual amount, all these studies find that R&D by the U.S. fertilizer industry is relatively small and represents a small share of industry revenue.

Table 4.2. Research and development (R&D) by the global fertilizer industry in 2006

Country or region	Dominant sector	R&D	Production value	Production
Leading countries		*Mil. U.S. dollars*		*Mil. Tons N,P,K nutrients*
China	State	22.0	21,525	45.9
U.S.	Private	19.1	9,152	20.0
Canada	Private	8.9	4,626	13.6
Russian Federation	Private	6.5	6,492	16.2
India	State	6.4	6,285	13.1
By region				
North America		28.1	13,778	33.5
Europe-ME		32.9	26,166	62.1
Asia-Pacific		35.2	33,188	70.3
Latin America		3.0	2,929	6.5
Global total		99.1	76,060	172.5

Source: USDA, Economic Research Service. Production of nitrogen (N), phosphorus (P_2O_5), and potassium (KCl) fertilizers are from Food and Agriculture Organization of the United Nations (FAO). Value of production is estimated by multiplying FAO production quantities by global fertilizer trade prices (nutrient basis) from Haver Analytics.

R&D is estimated as 0.25 percent of production value in Organisation for Economic Co-operation and Development and former Soviet Union countries and 0.125 percent of production value in developing countries using average global prices over 2002-05.

Table 4.3. Research and development (R&D) by the fertilizer industry in the United States

Year	Source	Industry R&D expenditures	
		Million nominal U.S. dollars	*Million constant U.S. dollars*
1975	Wilcke & Williamson (1977)	3.4	10.5
1978	Malstead, reported in Ruttan (1982)	3.0	7.7
1979	Malstead, reported in Ruttan (1982)	3.0	7.1
1984	Crosby (1987)	22.2	38.3
2006	Present study	19.1	19.1

Current expenditures adjusted for inflation by the U.S. Gross Domestic Product implicit price deflator (Economic Report of the President, 2009).

Source: USDA, Economic Research Service using studies listed in table.

References

Crosby, Edwin A. 1987. "Private Sector Agricultural Research in the United States," in V.W. Ruttan and C.E. Pray, (eds.), *Policy for Agricultural Research*, Boulder, CO: Westview, pp. 395-409.

Economic Report of the President. 2009. U.S. Government Printing Office.

Evenson, Robert E., and Larry E. Westphal. 1995. "Technological Change and Technology Strategy," in J. Behrman and T.N. Srinivasan (eds.), *Handbook of Development Economics, Vol. III*, Amsterdam: Elsevier Science, pp. 2209-2299.

Food and Agriculture Organization of the United Nations (FAO). FAOSTAT database, Rome, Italy, available at: http://faostat.fao.org/ (accessed October 2010).

Federal Reserve Bank of St. Louis. Producer Price Indexes, FRED Economic Data, available at: http://research.stlouisfed.org/fred2/categories/31 (accessed September 2009).

Haver Analytics. Commodity Prices, International Financial Statistics. DLX Databases, Haver Analytics, Inc., available at: www.haver.com/ (subscription required).

Heffer, Patrick, and Michel Prud'homme. 2008. *World Fertilizer and Fertilizer Demand, Global Fertilizer Supply and Trade, 2008-2009*, International Fertilizer Industry Association, Paris.

Hernandez, Manuel A., and Maximo Torero. 2011. *Fertilizer Market Situation: Market Structure, Consumption and Trade Patterns, and Pricing Behavior*, IFPRI Discussion Paper 01058, International Food Policy Research Institute, January.

Huang, Wen-Yuan, William McBride, and Utpal Vasavada. 2009. "Recent Volatility in U.S. Fertilizer Prices," *Amber Waves*, Vol. 7 Issue 1, U.S. Department of Agriculture, Economic Research Service, available at: www.ers.usda.gov/amberwaves/ march09/meatures/mertilizerprices.htm.

Intergovernmental Panel on Climate Change (IPCC). 2007. *Climate Change 2007: Mitigation of Climate Change,* Contribution of Working Group III to the Fourth Assessment Report of the Intergovernmental Panel on Climate Change, B. Metz, O.R. Davidson, P.R. Bosch, R. Dave and L.A. Meyer (eds.), Cambridge, UK: Cambridge University Press, available at www.ipcc-wg3.de/activity/assessment

International Fertilizer Industry Association (IFA). 2010. IFA Fertilizer Database, available at: www.fertilizer.org/ifa/ifadata/search (accessed April 2011).

International Fertilizer Development Center (IFDC). 2010. Virtual Fertilizer Research Center, IFDC, available at: www.ifdc.org/Alliances/VFRC.

International Plant Nutrition Institute (IPNI). IPNI Website, available at: www.ipni.net/ (accessed November 2009).

Nature. 2009. "The Shale Revolution: The Vast Reserves of US Natural Gas Must Be Used Judiciously To Ease the Transition to Clean Energy," Editorial, Vol. 460 (7255) July 30.

Park, Murray. 2001. *The Fertilizer Industry*, Cambridge, UK: Woodhead Publishing.

PotashCorp. 2007. *Overview of PotashCorp and Its Industry 2007*, PotashCorp, Saskatoon, Canada, available at: www.potashcorp.com.

PotashCorp. 2008. *Overview of PotashCorp and Its Industry 2008*, PotashCorp, Saskatoon, Canada, available at: www.potashcorp.com.

Ruttan, Vernon W. 1982. *Agricultural Research Policy*, Minneapolis: University of Minnesota Press.

Tomich, Thomas P., Peter Kilby, and Bruce Johnston. 1995. *Transforming Agrarian Economies: Opportunities Seized, Opportunities Missed*, Ithaca, NY: Cornell University Press.

U.S. Department of Agriculture, Economic Research Service (USDA/ERS). 2006. *Agricultural Resources and Environmental Indicators*, Economic Information Bulletin 16, available at: www.ers.usda.gov/publications/arei/ eib16/

Wilcke, H.G., and J.L. Williamson. 1977. *A Survey of U.S. Agricultural Research by Private Industry,* Agricultural Research Institute.

Yara International. 2007. *Financial Review 2007*, Oslo, Norway, available at: www.yara.com.

Farm Machinery Research and Development by Private Industry

David Schimmelpfennig and Keith O. Fuglie

Since the 19th century, private-sector firms and entrepreneurs have led the development of new farm machinery, with public institutions investing relatively little in this area. Private-sector makers of machinery capture the gains from their innovations through sales of their products, and the patent system protects and enforces their rights as innovators. As in other agricultural input sectors, large multinational corporations are engaged in developing and marketing farm machinery in global markets. While small- and medium-sized firms (as well as individual inventors and farmers) have historically been a source of innovation in farm machinery, their share of the global market appears to have significantly declined over the past decade. Consolidations and mergers have led to fewer and larger companies producing for global markets. The U.S.-based company John Deere is the world's largest manufacturer of farm machinery, with sales of over $16.5 billion in 2008. About 40 percent of these sales were to markets outside the United States and Canada. In the U.S. market, European and Asian firms are also major suppliers of farm machinery. Between 1994 and 2009, the market share of the four largest farm machinery manufacturers rose from 28 to 50 percent of total global sales (see table 1.7).

Global Farm Machinery Markets and Factors Affecting Farm Mechanization

Industry sources estimate that the global market for farm machinery exceeded $70 billion per year by 2005 (Freedonia, 2006). In real terms, sales of farm machinery are estimated to have grown by about 3 percent per year since 2000, with growth strongest in the Asia-Pacific region (table 5.1). Farm tractors make up the largest share, accounting for nearly 30 percent of global sales of new farm equipment. Harvesting and haying machinery make up another 22 percent of the global market. Equipment for such uses as planting, fertilizing, plowing, cultivating, irrigating, and spare parts accounted for the remainder.

The pattern and speed of farm mechanization is heavily influenced by relative scarcities of farm land and labor, the demand for labor from the nonfarm sector, and the aggregate demand for agricultural products (Hayami and Ruttan, 1985). Although mechanization may not directly lead to crop yield increases, it can contribute significantly to raising total factor productivity in agriculture by saving labor costs as well as reducing the resources required for maintaining draft animals. In developing countries, power-intensive operations like plowing, water pumping, grain milling, and transporting crops to markets tend to be mechanized first, with control-intensive operations like harvesting and crop husbandry shifting to mechanized techniques only when wages are high or rapidly rising (Binswanger, 1986). Mechanization also facilitates growth in farm size, and larger farms tend to adopt new forms of machinery much faster than small farms.

Table 5.1. The global market for agricultural machinery

		By type			By region			
Year	Total global farm machinery sales	Farm tractors	Harvesting & haying equipment	Other	North America	Europe, Africa & Middle East	Asia-Pacific	Latin America
	Million constant 2006 U.S. dollars							
1995	62,091	18,344	13,901	29,846	18,426	21,307	19,642	2,716
2000	61,397	18,713	12,778	29,907	16,855	22,824	18,719	2,999
2005	72,486	20,941	16,325	35,221	22,025	22,722	24,668	3,072
2010	82,858	24,307	18,018	40,533	22,450	23,994	32,658	3,756

Source: USDA, Economic Research Service using Freedonia (2006). Annual expenditures adjusted for inflation using the U.S. Gross Domestic Product implicit price deflator (Economic Report of the President, 2009).

Farm Machinery Market Structure, Innovation, and R&D Spending

The global farm machinery industry underwent significant structural changes during the latter half of the 20th century, with the largest firms growing their market share, primarily through mergers and acquisitions. The four largest firms increased their share of the global farm machinery market from about 28 percent in 1994 to 50 percent by 2009 (see table 1.7). By 2009, at least 10 companies worldwide had annual sales of farm machinery valued at over $1 billion; together, these companies accounted for about one-third of the global market.

While large firms account for most of the formal R&D by the farm machinery industry, small- and medium-sized firms play an outsized role in innovation. Evenson (1982), basing his analysis on the patterns of patent ownership for farm machinery innovations in the United States, characterized the farm machinery industry as one in which large firms have concentrated on making refinements and achieving economies of scale in the manufacture of innovations originating from small-sized entrepreneurs. Case studies from other countries have also demonstrated the important role of small, local entrepreneurs in developing adaptive innovations of farm machinery, such as in the emergence of the power tiller industry in Thailand (Wattanutchariya, 1983) and water pump set manufacturing in China (Huang et al., 2007). Inventive work on a particular operation may precede its widespread use by decades, and invention often reaches a peak during the initial adoption cycle when many small firms enter with alternative designs (Evenson, 1982). The most successful of these firms either grow or are bought up by larger firms that can offer scale economies in manufacturing and distribution. Recently, this pattern has been seen in the rapidly expanding demand for drip- and micro-irrigation technologies in response to growing water scarcity in some regions of the world. During 2006-08, both John Deere and an Indian firm, Jain Irrigation Systems, acquired a number of U.S. and foreign firms specializing in irrigation technology. By utilizing their global manufacturing and distribution networks, the two firms established themselves as global leaders in agricultural irrigation technology.

Fifteen companies worldwide had over $500 million in farm machinery sales in 2006 (table 5.2). The four leading farm machinery companies develop and produce multiline products, including tractors, harvesting equipment, and implements, whereas second-tier companies are more likely to specialize in certain types of machinery.

Three of the four leading manufacturers also produce nonfarm machinery, such as earth-moving and construction equipment or machines for home-gardening and lawn care. These companies report research intensities ranging between 2.1 and 3.7 percent of total farm and nonfarm machinery sales.

For the purpose of estimating farm machinery R&D, we apply this ratio to the companies' reported sales of farm machinery. Second-tier companies are characterized by a wider range of research intensities as well as less complete data available on R&D spending. But none of the firms on which we have data exceeded 4 percent as an R&D-to-sales ratio, and some had less than 1 percent.

For 11 firms from high-income countries for which we have data on farm machinery sales and R&D expenditures, the average research intensity was 2.40 percent. For six developing-country firms, the average research intensity was 0.82 percent. To derive an estimate of R&D spending by second-tier firms on which we lacked data, we multiplied their sales by the average R&D intensity ratio for this group of firms (2.40 percent of sales for firms from high-income countries and 0.82 percent of sales for firms from developing countries) to derive industry-level estimates of R&D spending.

Farm machinery sales and R&D expenditure estimates in 2006 are reported for different market segments in table 5.3. The top part of the table shows sales and R&D by size of firm: the totals for the four leading manufacturers and second-tier companies with at least $100 million in farm machinery sales. "Other manufacturers" include all nonlisted companies, and their total sales is simply the difference between the estimate of total global sales and the reported sales for first and second-tier companies in table 5.2. This category includes virtually hundreds of small and medium-sized companies, many with no formal R&D departments but which nevertheless are sources of innovation for the farm machinery industry. To get an estimate of R&D spending for this group, we assume an R&D-to-sales ratio of 0.82 percent (the average for second-tier firms from developing countries). Taken together, our estimate of total R&D by the global farm machinery industry in 2006 is $1.48 billion, with 57 percent of this attributed to the four leading farm machinery manufacturers. Geographically, firms based in the United States, the EU, and Japan are the global leaders in farm machinery sales and R&D spending. As in the case of other agricultural input industries, many of these firms produce machinery for the global market and locate manufacturing and R&D facilities in several countries. Among developing countries, India is an important manufacturer, especially of small four-wheel tractors, with several large (second-tier) companies that produce machinery for both the Indian and global market. China is also a leading manufacturer of farm machinery, although the Chinese farm machinery industry appears to be dominated more by small- and medium-sized firms.

Table 5.2. Companies with over $500 million in farm machinery sales in 2006

Company	Country of incorporation	Farm machinery sales	Total equipment sales	R&D / sales	Ag machinery product lines
		Mil. U.S. dollars		*Percent*	
Leading multiline farm machinery manufacturers					
Deere	U.S.	10,232	19,884	3.65	Multiline
CNH	Netherlands	7,809	12,115	3.03	Multiline
AGCO	U.S.	5,435	5,435	2.35	Multiline
Kubota	Japan	5,103	5,796	2.13	Multiline
Second-tier farm machinery manufacturers					
CLAAS	Germany	2,954	2,954	4.27	Harvesters, balers
Yanmar Co.	Japan	1,440	1,440	n.a.	Tractors
Iseki	Japan	1,391	1,391	2.60	Multiline
SAME Deutz-Fahr	Italy	1,303	1,303	2.50	Tractors, combines
Kuhn Group	Switzerland	976	2,622	3.13	Implements
ARGO Group Spa	Italy	n.a.	n.a.	n.a.	Multiline
Minsk Tractor Works	Belarus	937	937	n.a.	Tractors
First Tractor Co., Ltd.	China	630	769	1.01	Multiline
Kverneland ASA[1]	Norway	569	569	3.77	Multiline
Mahindra & Mahindra	India	575	2,086	1.22	Tractors, implements
TAFE	India	568	n.a.	n.a.	Tractors
Total for listed companies		39,921	57,300	70.3	
Global market total		73,579		6.5	

n.a. = not available.
[1]Kverneland ASA was acquired by Kuhn Group in 2009.
Source: USDA, Economic Research Service using corporate websites and annual financial reports.

Farm Machinery R&D in the United States

Estimates of expenditures by U.S. manufacturing companies on farm machinery R&D were reported periodically by the National Science Foundation from the 1960s through 1997.[1] In addition, data from at least two independent surveys of private agricultural research in the 1970s included estimates of farm machinery R&D.

Based on these sources and findings from this ERS study, private farm machinery R&D rose between the 1960s and 1970s, peaking at about $627 million (in 2006 U.S. dollars) in the mid-1970s, then fell below $400 million in the 1990s before recovering to over $500 million again by 2006 (table 5.4).

Part of the renewed growth in farm machinery R&D by U.S. firms may be due to several factors: growing global demand for labor-saving equipment; the development of precision agriculture with global positioning systems (GPS), yield monitors, and auto-steer guidance systems; and the need to meet stricter regulatory standards, such the U.S. Environmental Protection Agency's air pollution Tier 4 standards for off-road diesel equipment.

Table 5.3. Farm machinery research and development (R&D) by the private sector in 2006

Market segment	Companies	Farm machinery R&D	Farm machinery sales	Average R&D / sales
By company classification Leading multiline farm machinery	*Number*	*— Mil. U.S. dollars —*		*Percent*
companies	4	847	28,479	3.0
Second-tier farm machinery manufacturers	30	407	16,841	2.4
Other manufacturers (not listed)	n.a.	9	375	2.4
By region				
North America	n.a.	575	23,054	2.5
Europe-ME	n.a.	581	22,357	2.6
Asia-Pacific	n.a.	311	25,040	1.2
Latin America	n.a.	9	3,129	0.0
Global total, all manufacturers	n.a.	1,470	73,579	2.0

n.a. = not available.

Sources: USDA, Economic Research Service estimates using data from Freedonia (2006), company annual reports, interviews with company representatives.

Table 5.4. Farm machinery research and development (R&D) by U.S. companies

Year	Source[1]	Industry R&D expenditures	
		Mil. U.S. dollars	*Mil. constant 2006 U.S. dollars*
1960-64	National Science Foundation	73.0	395.4
1965-69	National Science Foundation	98.6	478.7
1970-74	National Science Foundation	104.6	396.9
1975-79	National Science Foundation	205.5	553.5
1975	Wilcke & Williamson (1977)	203.8	627.1
1978	Malstead, reported in Ruttan (1982)	225.0	575.1
1979	Malstead, reported in Ruttan (1982)	225.0	530.9
1981, 1983	National Science Foundation	284.0	534.7
1985, 1987,1989	National Science Foundation	377.2	601.3
1991, 1993	National Science Foundation	291.7	397.6
1995, 1997	National Science Foundation	280.0	347.6
2006	Present study	513.2	513.2

[1]The National Science Foundation stopped reported R&D for the U.S. farm machinery industry after 1997.

Current expenditures adjusted for inflation by the U.S. Gross Domestic Product implicit price deflator (Economic Report of the President, 2009).

Source: USDA, Economic Research Service using studies listed in the table.

References

Binswanger, Hans P.1986. "Agricultural Mechanization: A Comparative Historical Perspective," *The World Bank Research Observer* 1 (January): 27-56.

Economic Report of the President. 2009. U.S. Government Printing Office.

Evenson, Robert E. 1982. "Government Policy and Technological Progress in U.S. Agriculture," in Richard Nelson (ed.), *Government Support of Technological Progress: A Cross Industry Analysis*, New York: Pergamon Press.

Freedonia, 2006. *World Agricultural Equipment to 2010*, Report Number 2089, Freedonia Group, Cleveland, OH.

Huang, Qiuqiong, Scott Rozelle, and Dinghuan Hu. 2007. "Pump-Set Clusters in China: Explaining the Organization of the Industry That Revolutionized Asian Agriculture," *Asia-Pacific Development Journal* 14 (December): 75-105.

Hayami, Y., and Ruttan, V.W. (1985). *Agricultural Development: An International Perspective*, Baltimore: Johns Hopkins University Press.

National Science Foundation, Division of Science Resource Studies (various issues). *Research and Development in Industry*.

Ruttan, Vernon W. 1982. *Agricultural Research Policy*, Minneapolis: University of Minnesota Press.

Wattanutchariya, S. 1983. "Economic Analysis of the Farm Machinery Industry and Tractor Contractor Business in Thailand," in *Consequences of Small-Farm Mechanization*, Los Banos, Philippines: International Rice Research Institute, pp. 39-50.

Wilcke, H.G., and J.L. Williamson. 1977. *A Survey of U.S. Agricultural Research by Private Industry*, Agricultural Research Institute.

Private Research and Development for Animal Health

Paul Heisey and Keith O. Fuglie

Significant productivity gains in agriculture have been achieved through improvements in animal husbandry and health. The animal health industry has contributed in this area by developing and supplying new vaccines, medicated feeds, anti-infectives, paracitides, and other pharmaceuticals that have reduced animal mortality and morbidity and raised growth and reproductive rates. The animal health industry is research intensive—globally, it invests more than 8 percent of its net sales in R&D, with leading firms investing at even higher levels. The industry is a component of one of the world's largest and most research-intensive industries, the pharmaceutical industry, and several leading companies invest in both human and animal health. In recent years, animal health product sales have ranged from 2.5 to 3.0 percent of global pharmaceutical sales.

Researchers estimating private R&D for agricultural animal health face two major challenges: many of the leading pharmaceutical companies do not report animal R&D expenditures separately from total R&D spending (which, in most cases, is dominated by human health R&D); and second, even when reasonable estimates can be derived of company R&D spending for animal health, it may be impossible to distinguish between R&D for food animals versus nonfarm (companion and equine) animals. The approach in this study is to rely on company and industry information to estimate total animal health R&D expenditures and then apportion this R&D among food and nonfood animals according to species market shares of product sales.

In recent years, structural changes in the global pharmaceutical industry, including mergers, have affected the animal health industry. In 2009, the animal health industry underwent major structural realignment, with the number of first-tier companies (those with at least $1 billion in annual sales of animal health products) falling from eight to six. To assess the effects of recent mergers on concentration and R&D spending in the animal health industry, we have extended our data period to the end of 2009. We also report on some other mergers and acquisitions that took place in 2010 and 2011, but these involved second- and third-tier companies and are unlikely to significantly alter the level of concentration in the global animal health industry.

We find that globally, private-sector growth in animal health R&D over the past decade was mostly for nonfood animals. Spending on food-animal health R&D (in constant 2006 U.S. dollars) declined from over $900 million per year in the mid-1990s to under $800 million annually during 1999 to 2007 before recovering somewhat to $890 million/year over 2008-10. This trend reflects the stagnant market for food-animal health products. Mergers and acquisitions among major pharmaceutical companies have led to a relatively high degree of concentration in the global animal health market, with the four largest firms accounting for more than 50 percent of global sales in 2009.

Global Market for Animal Health Products

Global Sales of Animal Health Products

Various sources place the global market for animal health products between $16 billion and $18.5 billion in 2006.[1] While the overall market for animal health products rose from just $10.8 billion in 1991 to nearly $17.5 billion by 2009 (constant 2006 U.S. dollars), most of this growth was in the nonfood-animal market (fig. 6.1). The food-animal share of the global market for animal health products declined from 80 percent in the early 1990s to about 58 percent by 2009. Sales of animal health products for food-animal species have fluctuated somewhat over time but in constant dollars have ranged between $8 billion and $10 billion since the early 1990s.

Breakdown by Product Types and Species

The three principal product types that we include in the market for animal health products are pharmaceuticals (including anti-infectives and paracitides), biologicals (primarily vaccines but also including diagnostic products), and medicated animal feeds (primarily antibiotics).[2] Pharmaceuticals are by far the largest component of this market, with sales of about $11.7 billion in 2009, although sales of biologicals rose more rapidly between 2001 and 2009 (table 6.1). Medicated animal feed is the smallest segment of this market (sales of $2.2 billion in 2009) and has shown the slowest rate of growth over the past decade. The use of antibiotics in animal feed has come under increased scrutiny due to concerns about microbial resistance, and in the EU, its use in feed for growth promotion has been phased out.

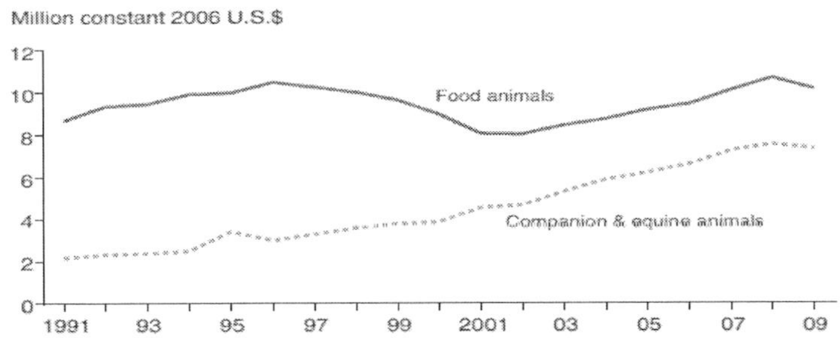

Source: USDA, Economic Research Service. Animal health product sales from Vetnosis, as reported in International Federation for Animal Health (various annual reports) and adjusted for inflation by the U.S. Gross Domestic Product implicit price deflator *(Economic Report of the President, 2009).*

Figure 6.1. Global sales of animal health products.

As previously mentioned, food animals account for just under 60 percent (about 44 percent in the United States) of the total market for animal health products. In 2009, animal health product sales for cattle (dairy and beef) made up about 25 percent of the total market, pigs nearly 18 percent, and poultry about 11 percent (table 6.1).

Table 6.1. Global market for animal health products

	Value of sales			Growth in sales over
By product type	2001	2006	2009	2001-09
	Mil. U.S. dollars			Percent per year
Pharmaceuticals	7,018	10,410	11,700	6.4
Biologicals	2,389	3,660	4,700	8.5
Medicated feeds	1,635	1,995	2,200	3.7
Total	11,050	16,065	18,600	6.5
By species	Market share			
	Percent			
Food Animals	64.0	58.9	58.1	
Cattle	28.2	27.2	25.3	
Sheep	5.1	4.8	4.3	
Pigs	17.9	16.1	17.7	
Poultry	12.8	10.8	11.3	
Nonfood animals & other	36.0	41.1	41.9	

Source: USDA, Economic Research Service using Vetnosis, as reported in International Federation for Animal Health (various annual reports).

Animal Health Industry Structure

The structure of the animal health industry can be described various ways, including by size of firm. In this chapter, first-tier animal health companies are those with at least $1 billion in animal health product sales in 2009, second-tier firms are those with between $300 million and $999 million in sales in 2009, and third-tier firms as those with less than $300 million in animal health product sales in 2009.

A fourth category, "biotechnology companies with animal health applications," comprises companies that mainly provide technology services to other firms in the industry. Industry structure may also distinguish between "discovery" firms and "generics" firms (the latter includes firms that do not develop their own products but produce off-patent products or products under license from developers).

Discovery firms tend to be much more research intensive than generics firms. All first-tier firms and most second-tier firms fall into the discovery firm category. A few firms specialize in products for food- or nonfood-animal species, although most of the larger firms produce products for both groups of species. Finally, industry structure may include a ranking of firms by their overall presence in the pharmaceutical industry.

First-Tier Animal Health Companies

Table 6.2 captures significant restructuring that occurred in the global animal health industry in 2009, when two first-tier companies acquired two other first-tier companies, further increasing concentration in the industry. Animal health is not the primary business of any of the first-tier companies listed in the table and, at most, contributes 10 percent of each firm's total pharmaceutical sales.

Over the past two decades, the largest firms set the pace in merger and acquisition activity in the animal health industry (fig. 6.2). In the 1990s, a number of other large pharmaceutical companies, among them Rhone-Poulenc, AkzoNobel, and Mallinckrodt, had substantial animal health sales. Rhone-Poulenc, a French chemical-pharmaceutical company, was a predecessor of Sanofi-Aventis. In 1997, this company and Merck formed a joint venture for their animal health segments, Merial.

In 2009, Merck sold its interest in Merial to Sanofi-Aventis, which then became the sole owner of Merial, and in 2009, Schering-Plough, another leading firm in the animal health industry, merged with Merck.

In 2007, Schering-Plough acquired another leading animal health company, Intervet. With the 2009 merger of Schering-Plough and Merck, Merck became the new parent company of the Intervet/ScheringPlough Animal Health subsidiary. In 2010, Sanofi-Aventis and Merck considered merging their animal health companies but later abandoned the idea.

Another major merger occurred in 2009 when Pfizer acquired Wyeth and merged its animal health business with Wyeth's Fort Dodge Animal Health subsidiary. In 2010, Pfizer acquired another second-tier animal health company, Alpharma.

By 2010, Pfizer's annual sales of animal health products exceeded $3.5 billion, making it by far the world's largest animal health company.

Second-Tier Companies

Unlike all of the first-tier animal health companies that are subsidiaries of large pharmaceutical companies, many second-tier companies specialize in animal health. Virbac, Ceva Sante Animale, and Vetoquinol are three independent animal health discovery companies with annual sales between $300 million and $1 billion. Lohmann Animal Health, which develops and sells poultry vaccines, may also be a second-tier company: in 2010, its parent company, the PH Wesjohann Group, reported $775 million (585 million euros) in animal health and nutrition sales, which includes nutritional feed products along with vaccines and other animal health products. We estimate that second-tier companies spent at least $218 million on animal health R&D in 2009 (about 14 percent of the private-sector total) and had a research intensity of 8.3 percent of sales, compared with 9.8 percent for first-tier companies (table 6.2).

Table 6.2. Major animal health companies

Company[1]	Animal health sales in 2006	Animal health sales in 2009	Animal health share of total pharmaceutical sales	Rank of parent firm in global pharmaceutical sales in 2010
1st tier companies (> $1 billion in animal health sales)	*Million U.S. dollars*		*Percent*	*Rank*
Pfizer	2,311	2,764	5.5	1
Intervet/Schering-Plough (parent: Merck)[3]	1,413	2,741	10.0	--[2]
Merial (parent: Sanofi-Aventis)[4]	2,195	2,554	6.3	6
Bayer	1,137	1,357	9.3	14
Elanco (parent: Eli Lilly)	876	1,207	5.5	11
Novartis	940	1,101	2.5	4
Fort Dodge (acquired by Pfizer from Wyeth in 2010)	936	--[2]	--[2]	--[2]
Schering-Plough (acquired by Merck in 2009)	910	--[2]	--[2]	--[2]
2nd tier companies ($300 -$999 million in animal health sales)				
Virbac	505	649	100.0	
Boehringer Ingelheim Vetmedica (BIV)	470	847	4.8	12
Ceva Sante Animale	301	413	100.0	
Alpharma Animal Health (acquired by Pfizer in 2010)	347	359	20.2	
Vetoquinol	266	350	100.0	
Lohmann Animal Health (parent: PH Wesjohann Group)	n.a.	n.a.	100.0	
Industry aggregate data for 2009	Companies	Animal health R&D	Animal health sales	R&D/Sales
	Number	*— Million U.S. dollars*	*—*	*Percent*
Total for 1st tier companies	6	1,149	11,724	9.8
Total for 2nd tier companies	5	206	2,490	8.3
Total for all others	n.a.	132	4,386	3.0
Global total - all animal health		1,487	18,600	8.0
Global total - food animal health		863	10,800	8.0

n.a. = not available.

[1]The first- and second- tier companies listed above are all classified as "discovery" companies except for Alpharma Animal Health, which is classified as a "generics" company.

[2]These companies had merged or been acquired by other firms by 2010.

[3]Intervet only in 2006 when it was a subsidiary of AkzoNobel. Schering-Plough acquired Intervet in 2007 to form Intervet/Schering-Plough. Merck acquired Intervet/Schering-Plough in 2009.

[4]Merial was a 50-50 joint venture between Merck and Sanofi-Aventis until 2009, when Merck sold its interest in Merial to Sanofi-Aventis.

Source: USDA, Economic Research Service. Animal health sales and R&D figures compiled from company financial reports and Animal Pharm Reports (2007); global ranking of pharmaceutical firms compiled from Fortune (2010).

Source: USDA, Economic Research Service using company websites.

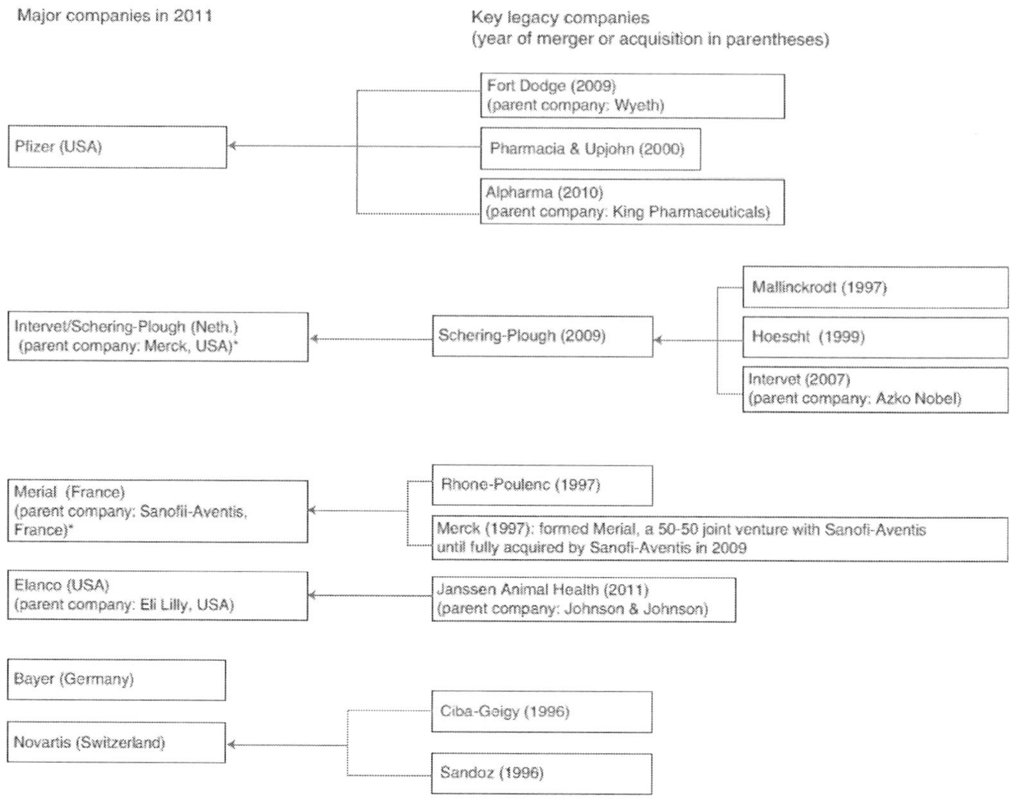

Source: USDA, Economic Research Service using company websites.

Figure 6.2. Evolution of major animal health companies.

Third-Tier Companies

Data are considerably more difficult to obtain for third-tier animal health companies, defined here as firms with annual global sales of less than $300 million. Using information from industry sources, we identified over 100 third-tier manufacturers of animal health care products and estimate their combined market sales to be $4.3 billion in 2006.[3] Of these companies, at least 30 are believed to invest in animal health R&D, although on average they spend less on R&D (as a percentage of sales) than either first- or second-tier firms. Based on observations from a limited number of firms in this category, we assume that third-tier firms spend an average of 3 percent of sales on R&D. Applying this research intensity to the total sales of these firms we derive an estimate of animal health research expenditures for this group of firms.

Biotechnology Companies with Animal Health Applications

In this study, we identified over 20 biotechnology companies that had animal health applications in 2006. Only four reported animal health as their main business. Most of these companies were relatively small, privately held, and appeared to be mainly technology service providers to other firms. Insufficient data made it difficult to estimate animal health R&D expenditures for these companies, so we omit these in our total R&D estimate for the animal health industry.

Concentration in the Global Animal Health Industry

The global pharmaceutical industry has been characterized by considerable merger and acquisition activity for at least the past two decades (CBO, 2006). Of the top 10 global pharmaceutical companies in 2008, only 2 have not been involved in significant horizontal merger activity.[4] Pharmaceutical company mergers or acquisitions have also affected the structure of the animal health industry, as all first-tier and several second- and third-tier companies are subsidiaries of large pharmaceutical companies. Thus, the principal reasons for the growing consolidation in the animal health industry are the factors driving consolidation in the larger pharmaceutical industry.[5] The data collected for this study show that concentration in the animal health industry has increased since 1994, with the global Herfindahl index rising from 510 to 827 between 1994 and 2009 (see table 1.7). With the recent mergers and acquisitions that occurred among the first-tier firms between 2006 and 2010, this ratio rose from 44 percent to just over 51 percent. This growth would place the animal health industry second to only the animal genetics industry in terms of four-firm concentration among the agricultural input industries considered in this report (see chapter 1 for more discussion).

Industry Investment in Animal Health R&D

To distinguish between the food and nonfood animal segments, we estimate R&D spending on food-animal health by apportioning total R&D according to the share of food animals in total product sales. This approach could over- or understate actual spending on agriculturally related R&D by the industry. Given the time lag between R&D spending and new product introductions, we would expect firms to allocate their R&D resources according to anticipated future market demand for new products. Since the market share of nonfood animal products has been rising over the last two decades, firms may be allocating a larger share of their current R&D to nonfarm-animal markets in response to this rising demand. On the other hand, many of the new products being introduced into nonfood-animal health markets are direct applications of human health care drugs that may not be relevant for food animals, such as treating the health conditions of aging companion animals. With a higher degree of "spillover" from human health R&D to the nonfood market segment, more of the animal-specific discovery R&D may in fact be directed to the unique problems faced by food animals. At present, we have insufficient information to determine which of these biases, if either, are significant, and so rely on this simple apportioning. Globally, private spending on animal health R&D increased from $806 million in 1994 to $1,449 million in 2010 (table 6.3). Spending on food animal health R&D grew at a substantially slower rate, rising from $645 million (or 80 percent of the total) in 1994 to $855 million (59 percent of the total) in 2010. In inflation-adjusted dollars, however, private R&D on food animal health appears to have declined since the mid-1990s, from over $900 million annually in 1995-96 to only $640 million per year by 2002, before recovering to around $800 million per year since 2006 (constant 2006 U.S. dollars). This lack of long-term growth in private-sector animal health R&D spending directly reflects the stagnant market for food-animal health products. The United States is the world's largest market for animal health products for both food and nonfood animals (with nonfood-animal health products accounting for about 56 percent of U.S. market sales). Companies based in the United States conduct about 42 percent of the

R&D by the global animal health industry (table 6.4). EU countries (the UK, Germany, the Netherlands, France, and Switzerland especially) account for 55 percent of global animal health R&D. Outside of these regions, companies based in China, Japan, India, Brazil, and Israel are also making significant investments in animal health R&D, but their combined share of global private animal health R&D is probably under 5 percent. U.S. and EU-based companies are also performing R&D for animal health markets in other regions of the world. Several of these companies have located research laboratories and have substantial product sales in these countries.

Table 6.3. Research and development (R&D) spending and research intensity by the global animal health industry

Year	Total private R&D spending for animal health	Private R&D spending for food animal health	Private R&D spending for food animal health	Animal health research intensity	Animal health research intensity by discovery firms
	Million nominal U.S. dollars		Million constant 2006 U.S. dollars	R&D/Sales (%)	
1994	829	664	858	8.6	11.8
1995	973	778	986	9.2	11.1
1996	985	767	953	9.1	9.6
1997	989	749	915	8.9	9.7
1998	977	720	870	8.7	9.8
1999	934	670	798	8.3	9.4
2000	937	655	763	8.5	9.7
2001	925	592	674	8.4	9.7
2002	933	590	662	8.2	10.2
2003	1,078	663	727	8.6	11.5
2004	1,190	712	760	8.7	11.0
2005	1,267	757	781	8.5	10.7
2006	1,349	794	794	8.4	10.4
2007	1,401	816	793	7.8	9.9
2008	1,638	960	913	8.5	11.6
2009	1,602	930	876	8.6	9.4
2010	1,595	941	878	8.6	9.8

Source: USDA, Economic Research Service. R&D spending estimated from company financial reports and as reported in Animal Pharm Reports (2003, 2005, 2006a, 2007). R&D spending for food animal health estimated by multiplying total R&D spending by the market share of food animals in total sales of animal health products. Market sales of animal health products from Vetnosis, as reported in International Federation for Animal Health (various annual reports). Annual expenditures converted into constant 2006 U.S. dollars using the U.S. Gross Domestic Product implicit price deflator *(Economic Report of the President, 2009).*

Our estimate of animal health R&D spending by U.S. companies, $546 million in 2006, compares favorably with industry estimates.

From its annual survey of member companies, P*h*RMA reports that R&D expenditures for veterinary pharmaceuticals were $496.3 million in 2006 (Pharmaceutical Research and Manufacturers of America, 2008).

Of this total, $356.4 million was spent in the United States and $139.9 million was spent by these firms abroad. Another U.S. industry group, the Animal Health Institute, reports that its member companies spent $663 million in animal health R&D in 2006 (Animal Health Institute, 2007).

The samples of companies included in these surveys are not identical and may vary from year to year, so these estimates are not directly comparable with ours.

Nonetheless, all three estimates show a similar trend in animal health R&D since 1994, with our estimate tracking somewhat closer to P*h*RMA's (fig. 6.3). We conjecture that most of the growth in animal health R&D spending in the United States since 2000 has been directed at nonfood-animal species.

Table 6.4. Private animal health research and development (R&D) by region in 2006

Sales and R&D by companies with HQ in:	Companies with sales > $50 million	R&D spending for animal health	Sales of animal health products by these companies	R&D/sales	Industry R&D share	Industry market share
	Number	*Million U.S. dollars*		*Percent*		
North America	8	562	6,145	9.1	42	38
Europe-Middle East	18	746	5,417	13.8	55	34
Asia-Pacific	19	39	2,676	3.0	3	17
Latin America	2	3	1,827	3.0	0	11
Global total	47	1,349	16,065	8.4	100	100

Notes: Sales and R&D figures include animal health for food and nonfood animals.

Source: USDA, Economic Research Service using company financial reports and Animal Pharm Reports (2007).

Market sales of animal health products from Animal Pharm Reports (2006c) and Vetnosis as reported in International Federation for Animal Health (various annual reports).

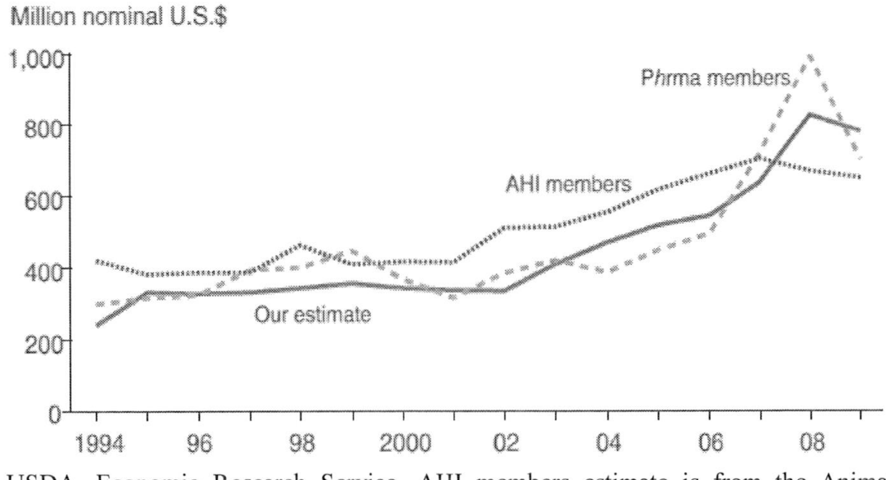

Source: USDA, Economic Research Service. AHI members estimate is from the Animal Health Institute annual survey of members. P*h*rma members estimate is from the P*h*rma annual survey of members. ERS estimate is derived from company financial reports for U.S.-based companies identified in our survey with investments in animal health R&D (see text).

Figure 6.3. Estimates of research and development spending by the U.S. animal health industry.

References

Animal Health Institute, 2007. Animal Health Institute News Release, Washington, DC, Oct. 3.

Animal Pharm Reports. 2003. *R&D in the Animal Health and Nutrition Industry,* London: Informa UK Ltd., February.

Animal Pharm Reports. 2005. *Animal Pharms' Top 20: 2005 Edition*, London: Informa UK Ltd., October.

Animal Pharm Reports. 2006a. *Animal Pharms' Top 20: 2006 Edition*, London: Informa UK Ltd., October.

Animal Pharm Reports. 2006b. *Generics in the Animal Health Industry*, London: Informa UK Ltd., May.

Animal Pharm Reports. 2006c. *World Animal Health Markets: 2006 Edition*, London: Informa UK Ltd., August.

Animal Pharm Reports. 2007. *Animal Pharms' Top Tier Companies: 2007 Edition*, London: Informa UK Ltd., September.

Congressional Budget Office. 2006. *Research and Development in the Pharmaceutical Industry*, Congress of the United States, October.

Danzon, Patricia M., Andrew Epstein, and Sean Nicholson. 2007. "Mergers and Acquisitions in the Pharmaceutical and Biotech Industries," *Managerial and Decision Economics* 28, 4-5 (June-August): 307-328.

Economic Report of the President. 2009. U.S. Government Printing Office.

Fortune. 2010. "Global 500: Our Annual Ranking of the World's Largest Corporations," CNNMoney.com.

International Federation for Animal Health. (various years). *IFAH Annual Report 2003 through IFAH Annual Report 2010*, Brussels, Belgium.

Pharmaceutical Research and Manufacturers of America. 2008. *Pharmaceutical Industry Profile 2008*, P*h*RMA, Washington, DC, March.

PhillipsMcDougall. Key Data, Animal Health, AgreWorld, data accessible from www.agreworld.com/ (subscription required).

ANIMAL GENETIC IMPROVEMENT BY THE PRIVATE SECTOR

Keith O. Fuglie and Paul Heisey

Selecting superior animals for breeding stock emerged as a specialized industry in the 1920s and 1930s for poultry, in the 1930s and 1940s for cattle, and in the 1950s and 1960s for hogs (Bogus, 1992; Willham, 1982; and Schneider, 2004). More recently, the private sector has also begun investing in aquaculture breeding, especially for salmonoids (salmon and trout species), shrimp (Pannaeus spp), and tilapia (tilapia spp). The genetic improvements and reproductive technologies developed by these firms have made significant contributions to raising the productivity of animal farming around the globe.

As is the case with crops, animal breeders (with the exception of dairy cattle) have been able to capitalize on the benefits from hybridization, or cross-breeding. The typical model that emerged in the poultry and swine genetics industries was one in which a company would invest in improving purebred lines (inbreds) and then sell hybrids (crosses between different purebred lines) to producers who would use these animals as production stock on their farms. The hybrids are often complex crosses involving four to six purebred lines. Companies would supply genetic material to farmers through "pyramid programs" (nucleus herds to multiplier herds to producer stock).

The animal genetics firms could maintain control over their intellectual property by guarding access to their purebred "nucleus" herds. Since the offspring from the hybrids produced on the farm would themselves breed inferior stock, farmers would return to the animal genetics companies to replace their production stock. This model served species with high fecundity rates, such as poultry and pigs, but was less suitable for cattle because of the greater expense and difficulty in maintaining viable inbred lines (Narrod and Fuglie, 2000). The reproductive technology of artificial insemination (A.I.), combined with expanded performance testing of progeny, was the basis of investment in dairy cattle breeding by both producer cooperatives and private companies. Beef cattle breeding has exploited cross-breeding for heterosis since the 1970s, but the dominance of uncontrolled mating under pasture and rangeland conditions across widely diverse production environments in many beef cow-calf operations, combined with relatively long generation intervals, has meant that, to date, large-scale private investment in beef breeding has been relatively limited.

More recently, new reproductive technologies, such as embryo transfer, and the tools of molecular biology have opened up new possibilities in the animal genetics industry. In dairy cattle, genomic evaluations using single nucleotide polymorphism (SNP) markers have been developed that increase the accuracy of genetic selection programs, lower the cost, and reduce the time necessary to evaluate A.I. sires (Strauss, 2010). Similar genomic evaluations are evolving for both the beef and swine industries, and specific SNP tests for various traits of economic importance, including genetic diseases and meat quality, have been widely available for several years. A.I. has been widely used in the cattle industry for many decades, but its use in swine breeding has been more limited because boar semen, unlike that of bulls, does not remain viable if frozen.

Over the last decade, however, technologies that extend the life of fresh boar semen have succeeded in significantly increasing the use of A.I. in swine breeding.

It is difficult to obtain reliable information on the animal genetics industry. Nearly all firms in the industry are privately held and do not publish information on company revenues or R&D expenditures.

To examine R&D spending and market structure in this sector, we contacted all of the major animal genetics companies in the poultry, swine, and cattle breeding sectors and a sample of aquaculture breeders and requested the relevant information. Most companies we contacted complied with our request, at least in part. This chapter is based on the information provided by these companies, information from company and industry websites, and interviews with knowledgeable persons from public- and private-sector animal breeding programs.

To maintain confidentiality, we report R&D and market sales figures only for the animal breeding sectors as a whole and not for individual companies.

Structure of the Animal Genetics Industry

Poultry

The poultry industry is composed of at least three distinct subsectors: the broiler and turkey industries for poultry meat and the layer industry for eggs.[1] Over the past two decades, the poultry breeding industry has undergone considerable consolidation, with a few companies dominating genetic supply in each subsector. Currently, broiler breeding is dominated by three firms and layer breeding and turkey breeding are each dominated by only two firms (fig. 7.1). The EW Group, Hendrix Genetics, and Groupe Grimaud, all European-based, have established themselves as global "multi-species" animal genetics companies. Their divisions may include broilers, layers, turkeys, other avian species, pigs, and aquaculture species. One U.S.-based firm, Cobb-Vantress (a subsidiary of Tyson Foods), specializes in broiler breeding. Cobb-Vantress, Aviagen Broilers (EW Group), and Hubbard (Groupe Grimaud) together supply at least 95 percent of the global commercial breeding stock for broilers.[2] Two companies, the EW Group[3] and Hendrix Genetics, supply nearly all the global breeding stock for turkeys and layers. In 2008, Groupe Grimaud established a new layer breeding subsidiary, Novogen, but its market share is not significant. In total, our survey identified 18 companies worldwide that appeared to be engaged in some poultry breeding. In addition to the four shown in figure 7.1 and Heritage Farms, six other companies were engaged in breeding specialty chickens for niche markets (e.g., free-range chickens, colored birds, ducks, geese, and other avian species), and seven served regional markets for broiler or layer breeding stock. These regional breeding firms, like India's VH Group and Thailand's CP Group, usually worked through joint ventures or licensing agreements with one or more of the major multinational companies to adapt breeding material for local markets.

A driving force behind consolidation in the poultry genetics industry has been economies of scale and scope in conducting research and marketing genetic products. The advent of molecular biology (marker-selected breeding, primarily) in poultry breeding has driven up the fixed costs of a breeding program while expanding the possibilities for genetic improvement. With larger fixed costs, higher returns can be achieved through larger scale. Firms have also achieved economies of scope by investing in multispecies breeding. These come about by sharing biotechnology research capacity as well as market distribution networks across species.

Swine

Until the emergence of dedicated swine breeding companies in the 1950s and 1960s, genetic improvement in pigs was conducted primarily through national breed registries. Organization of the breed registries was something of a cooperative venture and would often receive technical support from public research institutions. Individual breeders would register their prize purebred breeding stock through a registry and provide data to track performance of progenies. Breeding material might be exchanged or sold among members to improve the genetic performance of pedigree lines. Purebred offspring from superior sires and dams would be sold to farmers who would crossbreed these within their own breeding herd. In the 1950s, the first dedicated swine breeding companies began offering superior crossbreds to farm producers (Schneider, 2004). As with poultry breeding, these companies maintained their own purebred nucleus herds and developed hybrid crosses with superior performance.

Initially, these efforts focused on hybrid boars but later expanded their scope to include hybrid sows and gilts and, more recently, to boar semen administered through A.I.

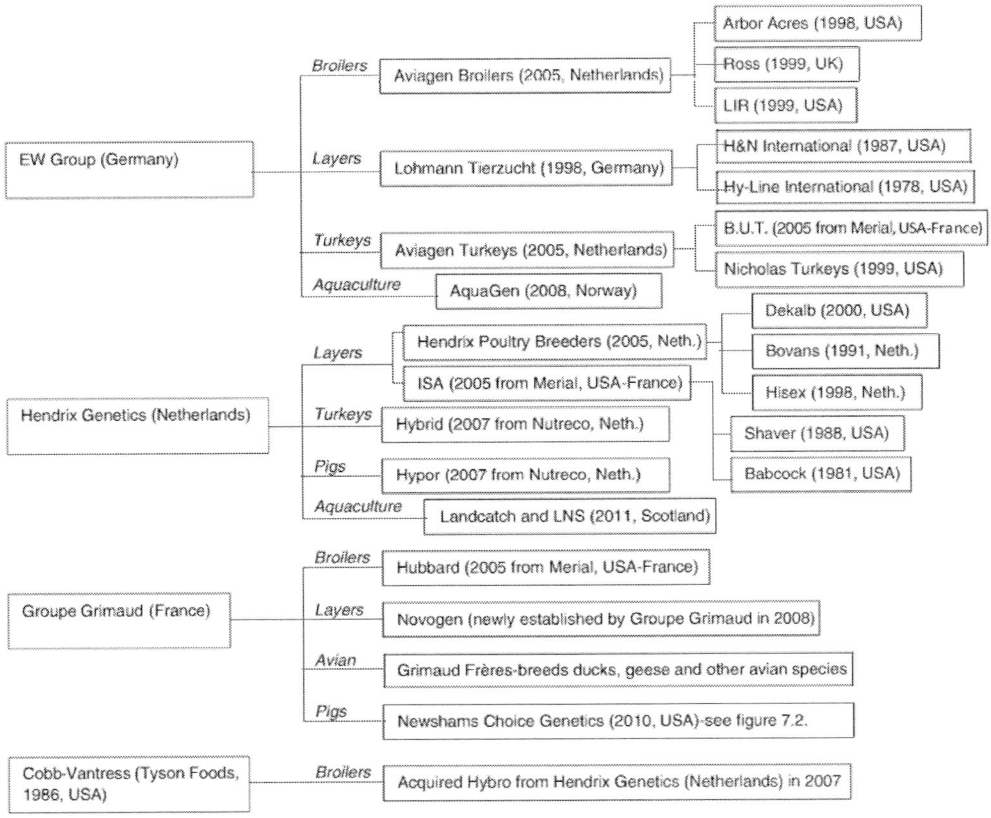

Name of company is followed by parent company, year of acquisition, and country of incorporation.

Figure 7.1. Evolution of the global poultry and multi-species genetics industry.

Over the last 15 years, the swine genetics industry has undergone significant consolidation although it remains considerably less concentrated than the poultry genetics industry (fig. 7.2). The leading global swine genetics company is PIC, which is owned by Genus, a publicly traded UK firm (Genus also invests in cattle breeding through its U.S.-based subsidiary, ABS Global). Another major swine breeding company is Smithfield Premium Genetics, a subsidiary of the vertically integrated pork producer and processor Smithfield Foods. This firm only supplies swine breeding stock internally to producers in the Smithfield system. Hypor (a subsidiary of Hendrix Genetics) and Newshams (a subsidiary of Groupe Grimaud) also have significant swine breeding R&D and international sales. Other important swine genetics suppliers include two farmer-owned cooperatives, Danbred (Danish-based) and TOPIGS (Netherlands-based). Farmer cooperatives in several countries supply swine breeding stock to their members, but Danbred and TOPIGS are unique in that they also export swine genetic material to other countries. A number of smaller firms, nearly all based in North America or Western Europe, invest in swine breeding.

The driving force behind consolidation in the swine genetics industry, much like in the poultry genetics industry, has been economies of scale in breeding and genetics. Unlike in poultry, however, independent pureline breeders working through national breed registers continue to play a role in animal genetic improvement and maintaining genetic diversity in commercial herds (Mabry, 2004).[4] Producer-owned farmer cooperative breeding programs are also active in some countries and in international markets.

Cattle

Genetic improvement in cattle continues to involve complex interactions between individual breeders, organized through breed associations; artificial insemination and embryo transfer companies; and public-sector research. Figure 7.3 depicts some of these relationships with a particular focus on artificial insemination companies and producer cooperatives. In the United States, for example, nearly all live breeding cattle in both dairy and beef production originate with breeders or seedstock producers whose revenues from sale of breeding stock are relatively small.[5] Out of nearly 800,000 beef cow-calf operations in the United States (USDA/NASS, 2009), between 70,000 and 75,000 operators, or around 10 percent, are associated with breed registries. Only about a quarter of these operators earn significant revenues from such sales, however, and a much smaller share earn annual revenues of up to $10 million per year from sales of live breeding animals, semen, and embryos (Marshall, 2010; Brester 2002). Similarly, out of some 70,000 dairy farmers in the United States, about 10,000 sometimes breed animals for sale as breeding stock, but only about 1,000 operators consider this activity as a significant source of revenue (Lawlor, 2010).

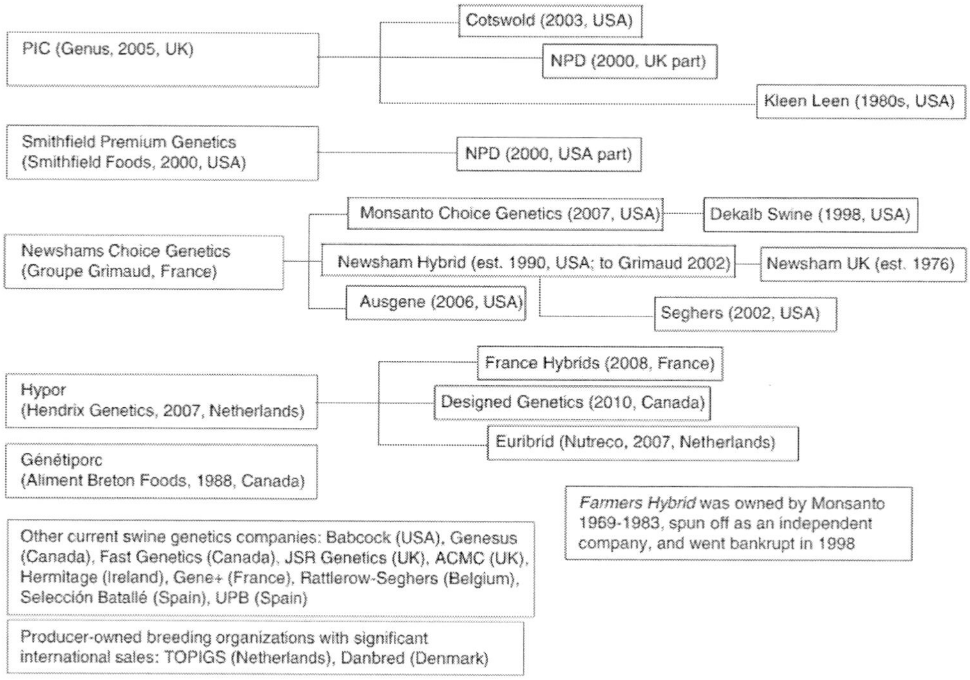

Name of company is followed by parent company, year of acquisition, and country of incorporation. PIC, Hypor, Newshams, Génétiporc, TOPIGS and Danbred appear to be the major multinational swine genetics companies. The other listed companies appear to serve primarily domestic markets.

Figure 7.2. Evolution of the global swine genetics industry.

Opportunities for cross-breeding and hybridization as a form of intellectual property protection have been more limited for cattle producers than for poultry and swine producers. In beef cattle, the relatively long gestation period (Narrod and Fuglie, 2000) combined with uncontrolled mating under rangeland conditions have limited the use of more complex cross-breeding schemes, although cross-breeding has made a significant impact in the beef industry. Today, 75-80 percent of U.S. beef cows are crossbred (Cundiff, 2007), and the supply of crossbred bulls or semen from crossbred bulls is the fastest growing sector of beef seedstock production (Marshall, 2010). The greater use of crossbred bulls and the development of new protocols for synchronizing heat in beef cattle may provide greater opportunities for more complex hybrid breeding schemes in beef cattle in the future.

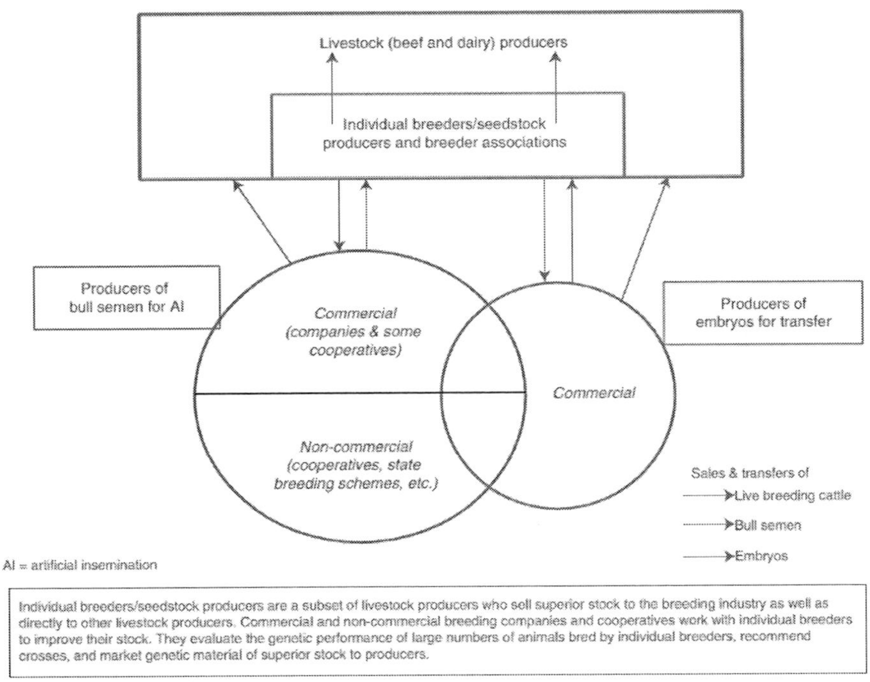

Figure 7.3. Private companies, producer cooperatives, and individual breeders in the global cattle genetics industry.

Dairy production, on the other hand, has been marked by the selection of superior animals from within the same breed, particularly in North America, where historically, the separation of dairy from beef production took place earlier than it did elsewhere in the world. Holsteins, deriving from a breed that originated in the Netherlands, now account for over 90 percent of U.S. dairy cows.[6] Artificial insemination, combined with large-scale performance recording schemes, has been the basis for investment in cattle genetics by producer cooperatives and companies since the 1940s. In the 1960s and 1970s, North American holstein semen was exported to Europe on a relatively small scale, but by the 1980s and early 1990s, bovine A.I. exports had developed into a large commercial market. In 1983, the International Bull Evaluation Service (Interbull) was founded with headquarters in Sweden, which helped to foster international genetic comparisons and the development of A.I.

companies outside of North America. By the late 1990s, some of these companies were significant competitors in the global market (Funk, 2006).

As with the poultry and swine industries, technological factors have strongly influenced cattle industry concentration; but unlike with poultry and swine, breeding programs for cattle have been generally unable to distinguish product lines. Each new technology is usually adopted by most major A.I. companies within a few years of being introduced, making price competition a major factor driving industry consolidation and globalization. A single bull can, theoretically, produce 50,000 offspring in 1 year through A.I., and techniques for freezing and storing semen, first developed in the early 1950s, facilitate the distribution of these progeny over a wide area.[7] Embryo transfer can allow more rapid dispersion of genes from elite females, although at nowhere near the rate A.I. allows for bulls. Embryo transfer technology, which was first established commercially in the 1970s, was adopted by A.I. companies in the 1980s.[8] Other technologies that have been adopted in the industry include genetic marker technology (1990s) and sexed semen (2000s) (Funk, 2006; Hassler, 2003; De Vries et al., 2008). More recently, single nucleotide polymorphism chips were introduced after having been developed through a collaboration involving a genetics-sequencing company whose primary focus is human health (Illumina), USDA, U.S. and Canadian universities, and the major North American A.I. companies. These chips potentially permit genetic evaluation of young sires much more quickly and cheaply than traditional progeny testing (Van Raden et al., 2009; Strauss, 2010).[9]

As a result of these patterns of technology diffusion, consolidation to achieve economies of scale and globalization have also occurred in cattle breeding, although not to the degree observed in poultry or swine breeding. The number of A.I. cooperatives and companies in the United States fell from about 200 in 1950 to approximately 20 in the 1980s (Narrod and Fuglie, 2000). In fact, by 1981, 11 companies provided 90 percent of the bovine semen processed in the United States; by the early 2000s, only 5 companies accounted for the same share. This group included three large cooperatives or cooperative alliances (Select Sires, Accelerated Genetics, and Genex/CRI), one privately held company (Alta Genetics, with headquarters in Canada and owned by a Netherlands-based holding company),[10] and one publicly traded company (ABS Global, which since 1999 has been part of Genus plc, based in the UK). Together with the Canadian cooperative alliance Semex (which has perhaps two-thirds to three-quarters market share in Canada), these companies constitute the six major A.I. organizations in North America. In Europe, cattle breeding organizations, whether cooperatives, companies, or government schemes, often hold market shares of 75-100 percent in their home countries, but no company has a European market share of over 25 percent (van Arendonk and Liinamo, 2003; Joint Research Council, 2007). Another notable European-based company marketing bovine semen globally, in addition to Genus, is CRV (Netherlands based, owned by Dutch and Flemish cooperatives), formerly known as Holland Genetics. Scandinavian-based Viking Genetics was founded in 2008 to consolidate the activities of cooperative-owned breeding programs in Denmark, Sweden, and Finland. In addition to these companies, members of the French association of A.I. cooperatives (UNCEIA) and alliances of German cooperatives and breeding associations (notably the Holstein breeding association DHV and cooperative unions TopQ and NOG) export significant amounts of bovine semen.[11] In 2009, UNCEIA, CRV, DHV along with VIT (German computing center for cattle data), and Viking Genetics announced they were forming a research consortium named EuroGenomics to aggregate the reference populations of Holstein bulls (with breeding values from progeny tests and known DNA profiles) used in their breeding programs. Finally, New Zealand-based LIC (cooperatively owned but with some shares publicly traded) has branches or agencies on several continents, in addition to its major position in the New Zealand market. In China, the recent expansion of the dairy industry has spurred the rapid growth of China Milk Products Limited (publicly traded on the Singapore Exchange). Although it operates only in the Chinese domestic market, by 2007, it had revenues from sales of semen and embryos, as well as physical quantities marketed, at a level comparable with those of some of the other companies mentioned in this chapter.

Aquaculture

The cultured fish industry has grown from only 11.3 million tons produced in 1985 to over 68 million tons by 2008 (FAO). A number of companies have emerged to supply superior genetic broodstock to aquaculture producers. Technological advances have enabled researchers to breed and multiply superior broodstock for a number of fish species, especially salmonoids like Atlantic salmon and rainbow trout and Penaeus species like whiteleg shrimp (*P. vannamei*). Although the fish-breeding industry is still in its infancy and most of the breeders are small, there may be significant opportunities for companies to enhance productivity through breeding and genetics and respond to the growing demands from aquaculture producers for superior, disease-free broodstock.

The emergence of a private aquaculture breeding industry is due in large part to earlier government-sponsored research that acquired, characterized, and improved fish genetic resources and established basic fish-breeding technologies. This has been especially apparent in Norway, which leads the world in breeding improved salmonoid (salmon and trout) species, and the United States, which has had notable successes in breeding shrimp. The modern Norwegian aquaculture industry dates to 1971, when the government established Akvaforsk (Institute for Aquaculture Research), with a mission to develop and transfer fish-breeding technology to the private sector. Breeding programs for Atlantic salmon and rainbow trout were established around this time. A number of salmonoid breeding companies emerged from this undertaking. One of the most successful is AquaGen, which was first established in 1992 as Norwegian Salmon Breeding AS (Norsk Lakseavl AS). In 2008, AquaGen was acquired by the German-based EW Group (see fig. 7.1).[12] Similarly, the shrimp (*Penaeus spp.*) breeding industry got its start from government-sponsored research in the United States. With financial support from USDA, the Oceanic Institute at Hawaii Pacific University successfully domesticated whiteleg shrimp (*P. vannamei*), which led to the development of genetically superior, specific-pathogen-free (SPF) broodstock. This technology was commercialized by such companies as Hi Health Aquaculture, Sygen, and Shrimp Improvement Systems (the latter, established in 1998, was acquired in 2007 by the CP Group, a Thai conglomerate).

Animal Genetics Biotechnology

A final component of the animal genetics industry is a group of biotechnology firms that either provide technology services to breeding companies or develop GM animals and fish. Most of these operations are small- or medium-sized firms that offer genomic services to companies in agriculture, health, and other life sciences sectors. A primary source of revenue for these companies is contract research with the larger animal-breeding companies. One company, AquaBounty, has developed a GM salmon that reportedly grows to harvest size in half the time it takes for conventional fish, and the company is currently seeking regulatory approval for this product. A number of other biotechnology companies are working on animal health technologies (see chapter 6 of this report). Regarding the use of biotechnology in the animal-breeding industry, our survey found wide-scale use of marker-selected breeding but little investment in transgenic animals or animal cloning. Concern about consumer acceptance of transgenic and cloned animals for food uses was cited by several companies as a reason for their lack of interest in pursuing applications of these technologies. For marker-selected breeding, however, companies perceive that this tool will speed up the rate of genetic progress in productivity and quality traits in animals and fish, and several companies have incorporated marker-selected breeding into their R&D programs.[13] Table 7.1 lists the major breeding companies in each animal and fish sector of the industry. In addition to the companies listed here, our survey identified 72 companies worldwide that appear to have some R&D investment in animal and fish breeding. However, the companies listed in table 7.1 account for most of the private R&D investments in this sector, especially for poultry, swine, and cattle. The aquaculture and animal biotechnology sectors, on the other hand, are composed mainly of small companies, including several others not identified in this study. But the aggregate R&D spending by these small companies is thought to be small.

The Market for Animal Genetics

As in the case of crop seed, farmers obtain their animal-breeding stock from diverse sources, including self-supply and other farmers, so determining the size of the commercial market for animal genetic material can be difficult. Moreover, the market may change over time. In developing countries, commercial companies may supply only a small share of total farm demand for breeding animals, but this share may increase over time as a country's livestock sector becomes more sophisticated and commercially oriented. In high-income countries, the increased use of A.I. in bovine and swine has reduced the number of live bulls and boars required to meet market demand for sire services as A.I. allows each animal to fertilize many more females. In hog production, the wider use of the "closed herd" system—in which a farm establishes its own nucleus herd of purebred parent lines to supply hybrid replacement gilts—has reduced the demand for replacement gilts from commercial seed stock companies. But use of this system may increase farm reliance on A.I. as a means of introducing improved genetics into their nucleus herd. In cattle production, even though the markets for live breeding animals—bulls and some replacement females—have a much higher monetary value than the markets for semen and embryos, practically all of the demand is met through individual breeders or small-scale breeding schemes.[14]

In the tables below, we have compiled information from several sources—estimates from companies and industry analysts as well as public statistics on animal production—to construct our own estimates of the size of commercial markets for animal genetics in the United States and for the world. These estimates only cover the portion of the animal genetics market supplied by commercial breeding companies and thus only cover a portion of total farm demand for animal breeding material. Moreover, given the lack of public data on these markets, these estimates are not definitive and subject to some margin of error. We estimate that sales of animal genetic material from commercial companies in 2006-07 were $4.06 billion globally and $1.35 billion in the United States (table 7-2).

Table 7.1. Major research and development (R&D) firms in the animal genetics industry

Sector	Major R&D firms and producer organizations[1]	Main R&D locations
Poultry – boilers	Cobb-Vantress (Tyson's Food, U.S.) Aviagen (EW Group, Germany) Hubbard (Groupe Grimaud, France)	U.S. UK, U.S. U.S., France, Brazil
Poultry – layers	Lohmann Tierzucht (EW Group, Germany) Hy-Line International (EW Group, Germany) ISA (Hendrix Genetics, Netherlands)	Germany U.S. Netherlands
Poultry – turkeys	Aviagen (EW Group, Germany) Hybrid (Hendrix Genetics, Netherlands)	U.S. Canada
Swine	PIC (Genus, UK) Smithfi eld Premium Genetics (Smithfi eld Foods, U.S.) Newshams Choice Genetics (Groupe Grimaud, France) Hypor (Hendrix Genetics, Netherlands) TOPIGS (producer cooperative, Netherlands) Danbred (producer cooperative, Denmark)	U.S., Canada U.S. U.S. Netherlands, Canada Netherlands Denmark, U.S.
Cattle (beef and dairy)	ABS Global (Genus, UK) Select Sires (producer cooperative, U.S.) Accelerated Genetics (producer cooperative, U.S.) CRI/Genex (producer cooperative, U.S.) Alta Genetics (Koepon Holding, Netherlands) Semex Alliance (producer cooperative, Canada) CRV (producer cooperative, Netherlands) Viking Genetics (producer cooperative, Scandinavian countries) LIC (producer cooperative, New Zealand)	U.S U.S U.S U.S Canada Canada Netherlands Denmark, Sweden, Finland New Zealand
Aquaculture	AquaGen (EW Group, Germany) Salmobreed (Norway) Landcatch Natural Selection (Hendrix Genetics, Netherlands) Troutlodge (U.S.) Genomar (Norway) Hi Health Aquaculture (U.S.) Shrimp Improvement Systems (CP Group, Thailand)	Norway – salmonoids Norway – salmonoids Scotland – salmonoids U.S. – salmonoids SE Asia – tilapia U.S. – Penaeus spp.[2] U.S. – Penaeus spp.[2]
Animal genetics biotechnology	Metamorphix (U.S.) AquaBounty (U.S.) Illumina (U.S.)	U.S. U.S. U.S.

[1]Company names are followed by corporate owner and country of incorporation in parentheses.
[2]The main *Penaeus* breeding species are *P. vannamei* (whiteleg shrimp) and *P. monodon* (tiger prawn).
Source: USDA, Economic Research Service survey.

The largest component of this was parent breeding stock for poultry hatcheries (broilers, layers, turkeys, and other fowl combined), at $1.74 billion for the world market and $362 million for the United States. Commercial penetration into the animal genetics market is highest for poultry, at nearly 100 percent in high-income countries and lower but increasing in developing countries.

For swine genetic material, we estimate commercial sales in 2006-07 to be $1.3 billion globally and $675 million in the United States. The largest component of this market is replacement gilts, and the fastest growing is A.I. According to the National Animal Health Monitoring Survey (NAHMS) conducted by USDA's APHIS,[15] the share of sows on farms with 100 or more head of hogs that were fertilized with A.I. rose from 1.1 percent in 1990 to 82.6 percent in 2006 (USDA/APHIS, 2005, 2008).

Table 7.2. U.S. and global markets for animal genetics in 2006/07

		Commercial sales of breeding materials[1]	
Sector	Types of breeding materials	United States	World
		Million U.S. dollars	
Poultry (broilers, layers, turkeys, and other fowl)	Female parent lines Male parent lines	362	1,742
Swine	Replacement sows & gilts Live boars Semen doses for A.I.	675	1,303
Cattle	Semen doses for A.I. Embryos	297	931
Aquaculture	Fish eggs and fry Prawn/shrimp larvae (broodstock)	12	87
Global total		1,346	4,062

n.a. = data not available. A.I. = artificial insemination.

[1]Commercial sales include the sale of animal genetic material supplied by genetics companies and producer organizations. In many countries, a share of breeding stock may be supplied by farmers themselves or through breeding associations. Thus, the sales estimates are generally below the total farm demand for animal genetic material.

Source: USDA, Economic Research Service.

Market penetration by commercial breeding companies is relatively low for cattle, and farmers tend to rely on self-supply or purchases of breeding stock from independent breeders or breeder associations. As a result, we do not include sales of live breeding animals for cattle and instead focus on sales of A.I. doses and embryos. We estimate global market sales of semen and embryos by commercial breeding companies to be $931 million per year in 2006-07. The total global market for these products was about $1.5 billion (the difference is made up by semen or embryos provided by producer cooperatives that do not market outside of their membership, government A.I. schemes, etc.). In figure 7.3., total global demand for bovine semen and embryos corresponds to the area of the two overlapping circles, while the portion supplied by commercial companies corresponds to the areas labeled "commercial." For the United States, commercial sales for bovine semen and embryos amounted to about $300 million in 2006-07. The use of A.I. was widespread in dairy cattle but was much less prevalent in beef cattle.[16] However, the number of embryo transfers for beef cows was roughly double the number for dairy cows.[17] During the last decade, embryo transfer grew rapidly in Latin America and Asia, grew more slowly in North America, and was relatively stable in Europe. As noted, however, one major impact of embryo transfer in dairy breeding has been the use of elite females in MOET herds to produce more bulls for progeny testing by the A.I. industry.

For aquaculture, we do not have sufficient information to estimate the size of the market for commercial breeding material, much of which has not been genetically improved. Some aquaculture producers continue to rely on wild seedstock and broodstock[18] rather than supplies from hatcheries, but the use of wild broodstock to restock fisheries is declining.

Research Spending in the Animal Genetics Industry

In total, the global animal genetics industry spent $295 million on R&D in 2006-07, or about 7.3 percent of sales (table 7.3). Nearly half of this amount is attributed to poultry breeding companies, where both sales and research intensity were relatively high (at least 8 percent of sales). Research intensity was lowest for the cattle sector, at less than 5 percent of sales.

While our survey identified 72 companies with at least some investments in animal genetic improvement, research spending tends to be concentrated among a few firms. In the poultry sector, four firms accounted for 97 percent of poultry R&D worldwide. For swine and cattle, the top four firms accounted for two-thirds of total industry R&D in both sectors. As previously mentioned, most of the companies in the animal genetics industry are privately owned firms. Our compilation also includes a number of producer-owned cooperatives that sell animal genetic material on a commercial basis to nonmembers. Table 7.4 shows total R&D spending, sales, and research intensity (R&D/sales) for publicly traded, privately owned, government-owned, and cooperatively owned companies. The research intensity of cooperatives was only about half the level of other firms, but this finding largely reflects the concentration of these firms in the cattle sector, where research intensity is lower than in other animal sectors. There may be a dichotomy of research intensities for privately owned companies. Research intensities of the larger privately held companies may be similar to those of publicly traded companies, but relatively small privately held companies may have lower research intensities.

Table 7.3. Research and development (R&D) spending and research intensity by the global animal health industry

Sector	Companies[1]	R&D expenditures	Breeding sales	R&D / sales
	Number	Million U.S. dollars		Percent
Poultry	18	141	1,742	8.1
Swine	16	96	1,303	7.4
Cattle (beef and dairy)	20	43	931	4.6
Aquaculture	17	6	87	6.5
Animal genetics biotechnology	5	10	n.a.	n.a.
Global total	72	295	4,062	7.3

n.a. = data not available.
[1]Companies that work on multiple species are counted only once in the total. Source: USDA, Economic Research Service survey.

In 2006, companies based in the United States and Canada had about 40 percent of the global market for animal genetic material and accounted for 50 percent of global R&D spending by this industry (table 7.5). European companies (Germany and the Netherlands are the leading countries) accounted for most of the rest of the R&D spending and about 57 percent of the global market. A number of foreign companies operate research stations in the United States. For example, UK-based Genus plc, the world's leading private cattle and swine breeding firm, locates its principal breeding stations for both commodities in the United States. In our survey, we asked companies to break down their R&D spending by the share spent in the United States and in other countries. When based on the location of R&D

facilities, the U.S. share of total industry spending on animal breeding and genetics research increased from 45 to 51 percent of the global total.

Table 7.4. Animal genetics firms and research and development (R&D) by type of company ownership in 2006/07

Type of ownership	Companies[1]	R&D expenditures	Breeding sales	R&D / sales
	Number	Million U.S. dollars		Percent
Publicly traded	6	62	642	9.7
Privately owned	54	195	2,518	7.7
Cooperatives	12	38	902	4.2
Global total	72	295	4,062	7.3

[1]Companies that work on multiple species are counted only once in the total. Source: USDA, Economic Research Service survey.

Table 7.5. The global animal genetics industry by region in 2006/07

Country	Companies[1]	R&D expenditures	Breeding sales	R&D/sales	Global R&D share	Global market share
	Number	Million U.S. dollars		Percent		
North America	36	147	1,615	9.1	50	40
Europe-ME	31	144	2,330	6.2	49	57
Asia-Pacific	5	5	117	4.1	2	3
Latin America2	0					
Global total	72	295	4,062	7.3	100	100

Note: Sales and research and development (R&D) expenditures are the totals for the companies incorporated in a particular country, including their sales and R&D in other countries.

[1]Companies that work on multiple species are counted only once in the total.

[2]Several multinational animal breeding companies operate research stations in Latin America, but we could find no major local breeding companies in this region.

Source: USDA, Economic Research Service survey.

Although our data do not show trends in animal genetics R&D spending over time, we can compare our findings with those of a study that used a similar methodology to estimate private-sector animal breeding research in 1996.

Narrod and Fuglie (2000) find that animal breeding companies (including foreign firms doing research in the United States) spent $144.5 million on R&D (in 2006 dollars) in the United States in 1996; our survey data showed expenditures of $155.8 million in 2006.

From 1996 to 2006, spending declined in U.S.-based poultry R&D but increased in swine and cattle R&D. Globally, total global breeding research on chickens (broilers and layers) increased by about 7 percent in real terms between 1996 and 2006 (from $122.9 million to $131.4 million), while swine breeding research more than doubled (from $37.9 million to $95.8 million).

For all animal genetics R&D, we estimate that global private R&D increased by 43 percent in real terms between 1996 and 2006.

Implications of Market Structure on Animal Breeding Research

The growing concentration in the market for animal genetics has raised concerns that these firms may hold excessive market power and reduce biodiversity (Gura, 2007). In a study of genetic diversity in commercial poultry flocks, Muir et al. (2008) found that commercial chicken breeds contained only about half the genetic diversity native to the species. The authors concluded that "these findings indicate that the poultry industry, across both the egg and meat pure-line stocks, has a narrowed genetic resource and possibly a reduced capacity to respond to future industry needs" (p. 17316). While the authors note that this reduction of diversity does not preclude future genetic progress, it does raise a concern that some traits, such as those conferring resistance to certain infectious diseases, may be lost through selective breeding in commercial poultry. It is not clear, however, that increased concentration in the industry has contributed to reduced biodiversity. Bugos (1992) reported that U.S. poultry breeding companies relied on a fairly narrow genetic pool as far back as the 1950s, when there was a larger number of companies engaged in commercial breeding. Similar concerns regarding loss of genetic diversity have been raised for the dairy industry (Young and Seykora, 1996; Hansen, 2000).

Among all sectors, levels of concentration in the animal genetics market are highest for poultry, lower for swine, lowest for cattle, and unknown for aquaculture. Given the absence of reliable information on the size of genetics markets and the prices of genetic materials, it is difficult to assess whether market concentration confers much market power to breeding companies. Moreover, compared with other agricultural input sectors, the size of the largest animal genetics companies are relatively small, certainly smaller in terms of total revenue than the largest crop seed companies. Genus (the one animal breeding company for which financial information

is publicly available and the global leader in swine and cattle genetics) had total net sales of $468 million in 2006, significantly below those of the largest six crop seed companies.

The share of sales devoted to R&D (research intensity) shows a similar pattern to the level of market concentration, with poultry at 8.1 percent of genetic sales, swine at 7.4 percent, and cattle at 4.6 percent. But market concentration is only one factor determining research intensity; others include projected market growth, technological opportunities (i.e., the ease at making genetic progress), the cost of research inputs, and the ability to appropriate gains from research (Pray and Fuglie, 2000). Several of these factors favor commercial poultry breeding, especially in broilers, and likely contribute to the higher research intensity of this sector (Narrod and Fuglie, 2000).

References

Bogus, Glenn E. 1992. "Intellectual Property Protection in the American Chicken-Breeding Industry," *Business History Review* 66: 127-168.

Brester, Gary W. 2002. "Meeting Consumer Demands With Genetics and Market Coordination: The Case of the Leachman Cattle Company," *Review of Agricultural Economics* 24,1: 251-265.

Cundiff, Larry. 2007. Leader (retired), Beef Genetics research program, Meat Animal Research Center, Agricultural Research Service (MARC/ARS), personal communication.

De Vries, A., M. Overton, J. Fetrow, K. Leslie, S. Eicker, and G. Rogers. 2008. "Exploring the Impact of Sexed Semen on the Structure of the Dairy Industry," *Journal of Dairy Science* 91: 847-856.

Estienne, M. 1993. *Swine Artificial Insemination Workshops. University of Maryland Swine Report*, University/Extension Factsheets and Research Report, Department of Agriculture, University of Maryland Eastern Shore.

Food and Agriculture Organization of the United Nations (FAO). Global Aquaculture Production Dataset, Rome, available at: www.fao.org/fishery/ statistics/global-aquaculture-production/query/en (accessed June, 2008).

Funk, Dennis A. 2006. "Major Advances in Globalization and Consolidation of the Artificial Insemination Industry," *Journal of Dairy Science* 89: 1362-1368.

Gollin, Douglas, Eric Van Dusen, and Harvey Blackburn. 2009. "Animal Genetic Resources Trade Flows," *Livestock Science* 120: 248-255.

Gura, Susanne. 2007. *Livestock Genetics Companies: Concentration and Proprietary Strategies of an Emerging Power in the Global Food Economy.* League for Pastoral Peoples and Endogenous Livestock Development, Ober-Ramstadt, Germany.

Hansen, L.B. 2000. "Consequences of Selection for Milk Yield from a Geneticist's Viewpoint," *Journal of Dairy Science* 83: 1145-1150.

Hassler, John F. 2003. "The Current Status and Future of Commercial Embryo Transfer in Cattle," *Animal Reproduction Science* 79: 245-264.

Johnson, Nancy L., and Vernon W. Ruttan. 1997. "The Diffusion of Livestock Breeding Technology in the U.S.: Observations on the Relationship Between Technical Change and Industry Structure," *Journal of Agribusiness* 15: 19-35.

Joint Research Council. 2007. *Consequences, Opportunities and Challenges of Modern Biotechnology for Europe*, Annex to Report 3, Case Studies Report, *The Impact of Primary Production Applications*, Office for Official Publications of the European Communities, European Commission, Luxembourg.

Lawlor, Tom. 2010. Director, Research and Development, Holstein Association USA, personal communication.

Mabry, John. 2004. "Evolution of the Independent Purebred Seedstock Industry in the USA," Proceedings of the 29th Annual National Swine Improvement Federation Conference and Meeting, December 9-10.

Marshall, Troy. 2010. Contributing Editor, *BEEF Cow/Calf Weekly*, personal communication.

Muir, W.M., G. Wong, Y. Zhang, J. Wang, M. Groenen, R. Crooijmans, H. Megens, H. Zhang, R. Okimoto, A. Vereijken, A. Jungerius, G. Albers, C. Taylor Lawley, M.E. Delany, S.A. Maceachern, and H.H. Cheng. 2008. "Genome-Wide Assessment of Worldwide Chicken SNP Genetic Diversity Indicates Significant Absence of Rare Alleles in Commercial Breeds," *Proceedings of the National Academy of Sciences* 105(45):17312-17317.

Narrod, Clare A., and Keith O. Fuglie. 2000. "Private Investments in Livestock Breeding With Implications for Public Research Policy," *Agribusiness* 16: 457-470.

Pray, Carl I., and Keith O. Fuglie. 2000. "The Private Sector and International Technology Transfer in Agriculture," in K.O. Fuglie and D.E. Schimmelpfennig (eds.), *Public-Private Collaboration in Agricultural Research: New Institutional Arrangements and Economic Implications*, Ames: Iowa State University Press, pp. 245-68.

Schneider, James F. 2004. "The Evolution of Breeding Companies as Seedstock Suppliers in the USA," Proceedings of the 29th Annual National Swine Improvement Federation Conference and Meeting, December 9-10.

Strauss, Stephen. 2010. "Biotech Breeding Goes Bovine," *Nature Biotechnology* 28(6): 540-543.

U.S. Department of Agriculture, Animal and Plant Health Inspection Service (USDA/APHIS). 2005. *Part IV: Changes in the U.S. Pork Industry, 1990- 2000*, USDA:APHIS:VS,CEAH, National Animal Health Monitoring System, #N428.0405

U.S. Department of Agriculture, Animal and Plant Health Inspection Service (USDA/APHIS). 2008. *Part IV: Changes in the U.S. Pork Industry, 1990- 2006*, USDA-APHIS-VS, CEAH, #N520.1108

U.S. Department of Agriculture, Animal and Plant Health Inspection Service (USDA/APHIS). 2009a. *Part IV: Reference of Dairy Cattle Health and Management Practices in the United States, 2007*, USDA: APHIS:VS, CEAH, #N494.0209

U.S. Department of Agriculture, Animal and Plant Health Inspection Service (USDA/APHIS). 2009b. *Beef 2007-08, Part II: Reference of Beef Cow-Calf Management Practices in the United States, 2007-08*. USDA: APHIS:VS, CEAH, #N512.0209.

U.S. Department of Agriculture, National Agricultural Statistics Service (USDA/NASS). 2009. *2007 Census of Agriculture*, available at: www. agcensus.usda.gov/publications/ 2007/index.asp.

Van Arendonk, Johan A.M., and Anna-Elisa Liinamo. 2003. "Dairy Cattle Production in Europe," *Theriogenology* 59: 563-569.

Van Raden, Paul M., George R. Wiggans, Curtis P. Van Tassell, Tad S. Sonstegaard, and Flavio S. Schenkel. 2009. "Benefits From Cooperation in Genomics," Interbull Annual Meeting Proceedings, Uppsala, Sweden, January 28-29.

Willham, R.L. 1982. "Genetic Improvement of Beef Cattle in the United States: Cattle, People and Their Interaction," *Journal of Animal Science* 54: 659-666.

Young, C.W., and A.J. Seykora. 1996. "Estimates of Inbreeding and Relationship Among Registered Holstein Females in the United States," *Journal of Dairy Science* 79: 502-505.

Private Research and Development for Animal Nutrition

Sun Ling Wang and Keith O. Fuglie

Processed or manufactured animal feed constitutes a major component of purchased farm inputs. In 2005, purchased feed accounted for 20 percent of total purchased inputs by U.S. farms and was equivalent to 22 percent of total livestock sales (USDA/ERS). Globally, the feed industry supplied 700 million tons of animal feed in 2008 (Best, 2009a). Like fertilizers, animal feeds are for the most part bulk products with few R&D inputs by manufacturing firms. However, specialty feeds and feed additives are high-value components of the feed market and are more technology intensive. Firms that supply these feeds have significantly higher R&D-to-sales ratios than manufacturers of bulk feeds.

Types of Manufactured Animal Feed

For optimal growth, farm animals require regular amounts of macro-ingredients (energy, primarily supplied by grains and grasses, and protein, mostly supplied by legumes) and micro-ingredients (vitamins and minerals) in their feed rations. In some cases, pharmaceuticals such as vaccines or antibiotics may be added to feed mixtures as well. Complete feed is industrial-compounded (blended) feed that fully matches the nutritional requirements of an animal and consists of both macro- and micro-ingredients. Premixes are ingredients used in the making of complete feeds and consist of protein-rich concentrates like soybean meal and/or micro-ingredients. Premixes may be sold to manufacturers of complete feeds or directly to farmers, who do their own blending with farm-grown or purchased grains to form complete feeds. In terms of total volume, complete feeds constitute 85-95 percent of total global feed sales while concentrates account for another 5-15 percent (Nutreco, 2008). Complete feeds and concentrates are high-volume bulk feeds. Specialty feeds and premixes of micro-ingredients, on the other hand, are low-volume (0.1 to 0.5 percent of total feed volume) but high-value products that may require significant investments in R&D to develop. We can distinguish between three main market segments for animal feed: (1) *compound feed* consisting primarily of complete feed and concentrates; (2) *nutritional feed additives* consisting of micro-ingredient premixes, such as vitamins and minerals, including enzymes, carotenoids, and amino acids; and (3) *medicated feeds*, which contain animal health pharmaceuticals. This chapter focuses on the first and second markets—compound feed and nutritional feed additives. Medicated feeds, which are manufactured primarily by pharmaceutical companies, were discussed in chapter 6.

Global Market for Manufactured Feed

According to industry sources (Best, 2009a), worldwide industrial production of manufactured animal feeds increased from just under 600 million tons in 1995 to 700 million tons in 2008 (fig. 8.1). Nearly all of this growth took place in developing countries. While the crop commodities used as raw ingredients by the feed industry are traded internationally, feed is manufactured primarily within the country in which it is consumed, due to the specific requirements of local markets and transportation costs. The largest manufacturer of animal feed is the United States, with about 25 percent of the global total, followed by the EU, China, and Brazil. The composition of manufactured feeds varies regionally. In the United States, a high proportion of manufactured feeds are concentrates due to the prevalence of onfarm feed mills in which premixes are blended with farm-grown grains; in the EU and developing countries, a higher proportion of the market is for complete feeds (Nutreco, 2008).

The size of the global animal feed market is difficult to determine because product quantity and price data are not readily available. However, due to the competitive nature of the market, compound feeds generally track movements in feed commodity prices, especially corn and soybean meal, the principal raw materials used in their manufacture. To assess the size of the global market for manufactured animal feed, we consider the compound feed and nutritional additive market segments separately. For compound feed, we estimate a global average price based on international trade prices for corn and soybean meal plus a markup for milling and distribution costs.[1] For nutritional feed additives, we rely on industry sources for the size of this market. In constant 2006 dollars, we estimate that global sales of compound feeds averaged around $100-120 billion between the mid-1990s and 2005 but then increased to over $220 billion by 2008 due to the sharp rises in commodity prices (see fig. 8.1).

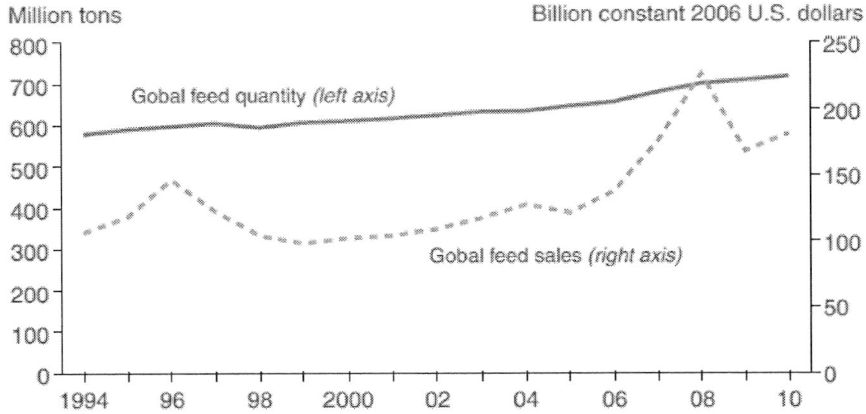

Source: USDA, Economic Research Service. Quantity of manufactured feed from Best (2009a). The average price of manufactured feed was derived by ERS based on a composite of international trade prices for corn and soymeal (IMF) plus a markup for processing and marketing costs. Nominal prices are adjusted for inflation using the U.S. Gross Domestic Product implicit price deflator (*Economic Report of the President*, 2009).

Figure 8.1. Global market for manufactured animal feed.

Nutritional feed additives are a relatively small but critical component of the animal feed market. This market consists of several unique segments, such as specific vitamins, amino acids, feed enzymes, and carotenoids, which provide animals with micro-nutrients as well as enhance digestion. Certain market segments may be dominated by a few companies, and with limited competition, there may be more scope for monopolistic behavior on the part of these firms. For example, in 1996, the U.S. company Archer Daniels Midland (ADM) and four Asian companies were found guilty of price-fixing behavior in lysine, an essential amino acid used in animal feed, in violation of U.S. anti-trust laws (Connor, 1997).

Generally, the level of market concentration in the manufacturing of bulk animal feeds is low, with the top 10 companies accounting for only 17 percent and the top 40 companies only 30 percent of global production in 2008 (Best, 2009b). However, in national or regional markets, and for specialty feed products, concentration ratios may be significantly higher.

R&D Spending by Animal Feed Companies

The U.S. agribusiness firm Cargill is the largest manufacturer of animal feed, followed by the Thai conglomerate Charoen Pokphrand (table 8.1). Four large European firms specialize in the production of nutritional feed additives, although some of the major producers of compound feeds, like Nutreco and ADM, manufacture both kinds of feed. R&D investments are made to develop new products, to reduce manufacturing costs through process innovations, and to determine optimal feed use in animal husbandry. The four firms listed in the table that specialize in nutritional feed additives have relatively high research intensities, at about 5 percent of sales. Of the largest producers of bulk feeds (some of which also produce specialty feeds and nutritional feed additives), we were only able to find sales and R&D figures for two. The average R&D-to-sales ratio for these firms was only 0.47 percent, which is similar to the findings of a 1975 survey of 31 animal feed companies in the United States that reported an R&D-to-sales ratio by these firms of 0.70 percent (Wilcke and Williamson, 1977). With such limited information, we can only make a rough estimate of R&D spending by this sector. However, like synthetic fertilizer, manufactured animal feed is largely a bulk agricultural input with relatively little R&D, so even an approximate estimate is not likely to distort the overall estimate of agriculturally related private-sector research. To estimate R&D spending by the animal nutrition industry, we use company R&D data when available and for other firms we apply representative R&D-to-sales ratios to firms in different segments of the industry. Moreover, we assume that only the 60 largest feed manufacturers worldwide conduct R&D. These firms accounted for about 30 percent of global production in 2006. For firms in high-income countries, we assume an R&D-to-sales ratio of 4.7 percent for manufacturers of nutritional feed additives and 0.50 percent for producers of bulk feeds. These are average R&D intensity ratios observed from eight feed manufacturers for which we have data, and the ratios are close to those reported by Wilcke and Williamson (1977) in their 1975 survey of U.S. agricultural input producers. For firms in developing countries, all of which produce primarily bulk feed products, we assume half this level, or 0.25 percent of sales. Evenson and Westphal (1995, table 37.1, p 2242-3) show that average R&D intensities of manufacturing industries in developing countries are typically half or less the average level for high-income countries. To estimate sales for the top 60 producers in the industry, we apply our estimate of the average global wholesale price of manufactured feed over 2000-2005 (in constant 2006 U.S. dollars) to the production volumes for 2006 reported by Best (2009b). We use the 2000-2005 average price ($199/ton) rather than the 2006 price of feed to avoid distortions caused by the inflated feed prices during 2006-08. Firms are unlikely to change their R&D expenditures quickly in response to price fluctuations.

Total R&D spending on animal feed by the largest 60 feed manufacturers was $375 million in 2006 according to our estimates (table 8.2). Companies located in the Europe-Middle East region made up 62 percent of the total, with Dutch and German firms both ahead of U.S. firms. Relatively high expenditures on animal feed R&D by European firms may be attributed to stricter EU regulations on the use of antibiotics, hormones, and animal parts in animal feed products. Such regulations increase farm demand for alternative feed ingredients and husbandry methods to provide for animal health and growth. It is likely that at least half of the total R&D by the animal feed industry is conducted for the nutritional feed additive segment of the market. The four companies listed in table 8.1 that specialize in these feeds alone spent $163 million on R&D, or 43 percent of our estimate of total R&D by the feed industry. Applying our method for estimating R&D for other firms in this market segment raised the total for R&D spending on nutritional feed additives to $215 million, or 57 percent of the feed industry total.

A comparison of our estimate for 2006 with estimates from five other studies from 1975-96 reveals a declining trend in R&D spending by the U.S. animal feed industry (table 8.3). The 1975 survey by Wilcke and Williamson (1977) found $85 million (in 2006 dollars) in animal feed R&D in the United States that year. Our estimate for 2006 was $64.5 million. Based on these study findings, real R&D spending on feed in the United States declined by an estimated 25 percent over the past three decades.

Table 8.1. Major manufacturers of animal feed in 2006

		Figures for animal feed business segment in 2006			
Company	Country	Production	R&D	Sales	R&D/sales
Major producers of bulk feeds (compound and premix concentrates)[1]		*Million tons*	*Million U.S. dollars*		*Percent*
Cargill/Agribrands	U.S.	17.5			
Charoen Pokphrand	Thailand	15.2			
Land o' Lakes Purina	U.S.	11.5	11.9	2,711	0.44
Tyson Foods	U.S.	10.0			
Zen-Noh Cooperative	Japan	7.8			
Nutreco	Netherlands	6.1	19.1	3,808	0.50
Ucaab Cooperative	France	4.0			
AG Abri	UK	3.8			
Smithfield	U.S.	3.6			
Sadia[2]	Brazil	3.5			
Provimi	Netherlands	3.3			
Hope Group[3]	China	3.2			
Archer Daniels Midland (ADM)	U.S.	3.2			
Ridley	Australia	3.2			
Perdigao[2]	Brazil	3.0			
Major firms specializing in nutritional feed additives					
DSM	Netherlands	77.9	1,371	DSM	Netherlands
BASF	Germany	29.2	858	BASF	Germany
Degussa[4]	Germany	31.4	651	Degussa[4]	Germany
Adisseo[5]	France	25.1	632	Adisseo[5]	France

[1]Some of these firms may also produce specialty feeds and nutritional feed additives.

[2]Sadia and Perdigao merged to form Brasil Foods in 2009.

[3]Hope Group includes New Hope Group and East Hope Group.

[4]Degussa was acquired by Evonik Industries, also a German firm, in 2007.

[5]Adisseo was acquired by CNCC, a Chinese firm, in 2006.

Sources: USDA, Economic Research Service using feed production estimates from Best (2006) and estimates of feed sales and research and development (R&D) from company annual reports.

Table 8.2. Research and development (R&D) spending by the animal nutrition industry in 2006

Sales and R&D by companies with their headquarters in:	Companies	Animal nutrition R&D	Animal nutrition sales	R&D/sales
	Number	*Million U.S. dollars*		*Percent*
North America	21	66	11,803	0.56
Europe-ME	32	232	17,036	1.36
Asia-Pacific	25	71	11,931	0.59
Latin America	6	7	3,101	0.22
Global total	84	375	141,770	0.26

Source: USDA, Economic Research Service estimates: Animal nutrition R&D from company financial reports where available or estimated by applying representative R&D/Sales ratios to firms in developed and developing countries; animal nutrition sales estimated from company financial reports where available or estimated by multiplying a representative feed price to feed production statistics for major firms given in Best (2009b).

Table 8.3. Private animal nutrition research and development (R&D) in the United States

Year	Source	Industry R&D expenditures	
		Million U.S. dollars	*Million constant 2006 U.S. dollars*
1975	Wilcke and Williamson (1977)	27.6	85.0
1978	Malstead, reported in Ruttan (1982)	30.0	76.7
1979	Malstead, reported in Ruttan (1982)	33.0	77.9
1984	Crosby (1987)	42.5	73.4
1996	Fuglie et al. (2000)	48.5	60.3
2006	Present study	64.5	64.5

Current expenditures adjusted for inflation by the U.S. Gross Domestic Product implicit price deflator (*Economic Report of the President*, 2009).

Source: USDA, Economic Research Service using data from studies in table.

References

Animal Pharm Reports. 2006. *Animal Pharm's Top 20: 2006 Edition*, Informa UK Ltd., London.

Ash, Mark, William Lin, and Mae Dean Johnson. 1984. *The U.S. Feed Manufacturing Industry*, Statistical Bulletin Number 768, U.S. Department of Agriculture, Economic Research Service.

Best, Peter. 2009a. "World Feed Panorama," *Feed International*, January, pp.12-15.

Best, Peter. 2009b. "Top Companies," *Feed International*, October, pp. 12-14

Connor, John M. 1997. "The Global Lysine Price-Fixing Conspiracy of 1992- 1995," *Review of Agricultural Economics* 19 (2):412-427.

Crosby, Edwin A. 1987. "Private Sector Agricultural Research in the United States," in V.W. Ruttan and C.E. Pray (eds.), *Policy for Agricultural Research*, Boulder, CO: Westview, pp. 395-409.

Economic Report of the President. 2009. U.S. Government Printing Office.

Evenson, Robert E., and L.E. Westphal. 1995. "Technological Change and Technology Strategy," in J. Behrman and T.N. Srinivasan (eds.), *Handbook of Development Economics, Volume III*, Amsterdam: Elsevier Science.

Fuglie, Keith O., Clare A. Narrod, and Catherine Neumeyer. 2000. "Public and Private Investments in Animal Research," in Keith O. Fuglie and David E. Schimmelpfennig (eds.), *Public-Private Collaboration In Agricultural Research*, Ames: Iowa State University Press, pp. 117-151.

Gill, Clayton. 2006. "Top 25 Companies," *Feed International*, October/ November, p. 10.

International Monetary Fund (IMF). Primary Commodity Prices, available at: www.imf.org/external/np/res/commod/index.asp (accessed November, 2009).

Nutreco. 2008. *Annual Report 2007*, Amsterdam.

Ruttan, Vernon W. 1982. *Agricultural Research Policy*, Minneapolis: University of Minnesota Press.

U.S. Department of Agriculture, Economic Research Service (USDA/ERS). Data Sets, Farm Income, available at: www.ers.usda.gov/data/farmincome/ (accessed January, 2009).

Wilcke, H.G., and J.L. Williamson. 1977. *A Survey of U.S. Agricultural Research by Private Industry,* Agricultural Research Institute.

RESEARCH AND DEVELOPMENT IN THE FOOD MANUFACTURING INDUSTRY

Kelly Day-Rubenstein and Keith O. Fuglie

Food manufacturing and processing companies produce intermediate foodstuffs or edible products for human and animal consumption.[1] The process of turning raw agricultural outputs into food, beverage, and tobacco products adds significant value to agricultural raw commodities (Gopinath and Roe, 1996); in 2009, it accounted for 13.5 percent of total U.S. manufacturing in terms of shipments (Bureau of the Census, 2010). Among all components of U.S. food and beverage manufacturing, meat processing is the largest (about a quarter of total shipments in 2009), followed by beverages, dairy, other food products, grains and oilseeds, and fruits and vegetables.

The food manufacturing industry differs significantly from the input industries reviewed in this report. Its work generally lies in post-harvest processing. While some firms in this sector do invest in raising farm productivity—plantation companies and animal feed manufacturers, for example— we have tried to include those investments within the respective farm input industries described elsewhere in this report. While most of the investments in R&D and innovation described in this chapter do not directly affect farm productivity, we assess them to make our estimates comparable with those of other studies. Klotz et al. (1995) report separate estimates of private-sector R&D spending in the United States by agricultural input industries and food manufacturing, and Alston et al. (2010) report combined estimates for private-sector food and agricultural R&D. Previous global estimates of private R&D spending have also lumped the food and agricultural input industries together (James, 1997; Pardey et al., 2006). By including food manufacturing R&D in our survey, we can compare our estimates with those of other studies (including USDA/ERS, 2010b) and at the same time provide richer detail about the share of the total directed at raising agricultural productivity or post-harvest processing.

R&D Spending by the Food Manufacturing Industry

The food manufacturing industry encompasses operations ranging from small processing firms to large multinational corporations. Large companies (defined here as those with annual revenue or turnover[2] in excess of $1 billion) account for a significant portion of this industry. Unfortunately, many of these companies do not make their R&D investments public (see table 9.1). Cargill, the largest company in this sector, is privately held and does not release R&D data. Others, such as Coca-Cola, consider R&D investment to be confidential business information.

The leading companies in this sector—Cargill, Nestlé, ADM, and Unilever— all have annual turnover or revenues in excess of $50 billion. R&D expenditures as a percentage of sales vary significantly among the largest companies that make this information available. Nestlé and Unilever invested more than $1 billion in companywide R&D in 2008, and Unilever's ratio of R&D to sales was over 2 percent. Sysco conducted no R&D at all.

Many food manufacturing companies often operate in other sectors, which further limits the usefulness of company data. For example, a considerable portion of Unilever's sales come from home and personal care products, and the firm's R&D spending includes investments in these areas as well as in food manufacturing. Nestlé produces pharmaceutical products. Few of these companies parse out their research spending by division or sector. Some companies conduct research that is directly related to agriculture, such as poultry breeding by Tyson Food (through its subsidiary, Cobb-Vantress) and tomato and pepper breeding by Campbell, as well as research related to new product development and process innovation.

Table 9.1. Sales and research and development (R&D) for leading food manufacturing companies in 2008

Company	Country	Net sales	R&D	R&D/sales	Significant agricultural R&D
		Million U.S. dollars		*Percent*	
Cargill	U.S.	120,400	n.a.		
Nestlé	Switzerland	91,896	1,653	1.80	Cocoa, coffee
Archer Daniel Midlands	U.S.	69,816	49	0.07	
Unilever	Netherlands	56,941	1,277	2.24	Tea
Pepsi	U.S.	43,251	282	0.65	
Kraft Foods	U.S.	42,201	499	1.18	
Sysco	U.S.	37,552	0	0.00	
Coca-Cola	U.S.	31,944	n.a.		
Wilmar International	Singapore	29,145	n.a.		Palm oil
Tyson Foods	U.S.	28,130	n.a.		Poultry
Smithfield	U.S.	14,264	91	0.64	Swine
Conagra	U.S.	13,809	69	0.50	
General Mills	U.S.	13,652	205	1.50	
Sara Lee	U.S.	13,450	n.a.		
Kellogg	U.S.	12,822	181	1.41	
Dean Foods	U.S.	12,455	8	0.06	
Land O Lakes	U.S.	12,039	40	0.33	Forage, dairy
Sime Darby	Malaysia	10,894	n.a.		Palm oil, rubber
Heinz	U.S.	10,071	n.a.		Tomato
Bunge	U.S.	10,028	34	0.34	
Cadbury	UK	9,960	128	1.28	
Campbell	U.S.	8,391	115	1.37	Tomato, pepper
Dole	U.S.	7,732	n.a.		Fruit

n.a. = not available.

Source: USDA, Economic Research Service using company annual reports and Fortune, May 8, 2008.

OECD Data on R&D in the Food Manufacturing Sector

The Organisation for Economic Co-Operation and Development produces the Business Expenditure on R&D (BERD) database (part of the Structural Analysis Statistics, or STAN database). The database provides the most comprehensive assessment available of R&D by the food manufacturing industry in high-income countries as well as in a few other countries.[3] All data are at an industry level. The BERD data generally use an enterprise-based approach: the R&D of a given enterprise will be classified by its primary industry only. This, combined with missing information, may lead to understating (or overstating) the R&D investment by a particular industry. Changes in classification schemes can also affect R&D data, though this has been less of an issue with food manufacturing industries.[4] The OECD data combine the food, beverage, and tobacco industries (International Standard Industrial Classifi cation codes 15 and 16).

Data are not available for every year for every country, so we report them here as an annual average within 5-year intervals that cover 1990-94, 1995- 99, 2000-04, and 2005-07 (see table 9.2). According to the OECD, R&D spending by the U.S. food, beverage, and tobacco manufacturing industry averaged $2.37 billion per year during 2005-07 (constant 2006 U.S. dollars), which was the highest amount globally, closely followed by Japan at $1.99 billion per year. Globally, R&D expenditures in food manufacturing have increased over time. From 1990-94 to 2005-07, global food industry R&D increased from $7.4 billion to $8.2 billion annually (in constant 2006 U.S. dollars). R&D expenditures also grew as a percentage of food industry value added (GDP), from 0.9 percent of GDP in 1990-94 to 1.6 percent of GDP in 2005-07 (table 9.2).

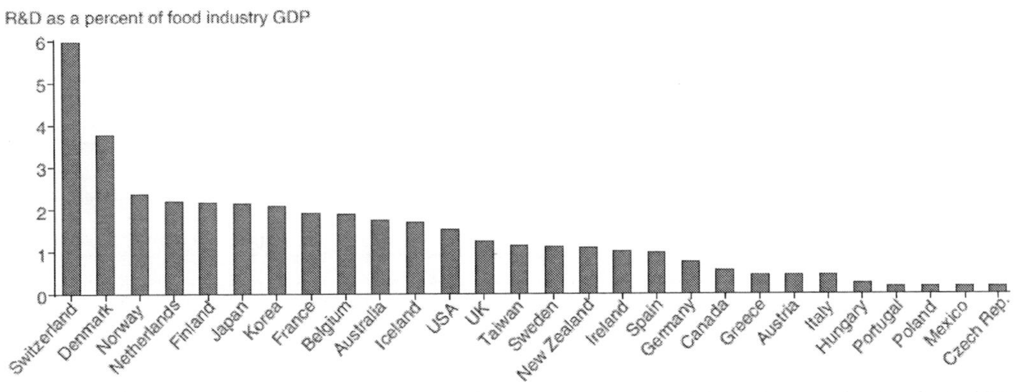

The figures show the average research and development (R&D)/Gross Domestic Product (GDP) percentage over 2000-2007.

Source: USDA, Economic Research Service using data from Organisation for Economic Co-operation and Development (OECD).

Figure 9.1. Research intensity in food manufacturing in OECD countries.

Research intensity varies considerably across countries (fig. 9.1). For the United States, research intensity was 1.53 percent during 2000-2007, about the average for all OECD countries. Among OECD countries, Switzerland, Denmark, Norway, and the Netherlands have the highest research intensity in food manufacturing. This partly reflects the presence of large, multinational, and R&D-intensive food companies in these countries, such as Nestlé (Switzerland) and Unilever (Netherlands) (see table 9.1).

These companies likely dominate national totals for these countries, even though some of the R&D by these companies may be conducted outside their home country. Generally, research intensity in the food industry is considerably less than that in manufacturing industries as a whole. For the 2000-2007 period, research intensity in all manufacturing industries among OECD countries was 7.6 percent, compared with 1.6 percent in the food manufacturing industry (OECD, 2010).

Table 9.2. Food Industry research and development (R&D) expenditures by country and region

	Average annual food R&D expenditure				Food industry R&D/GDP			
Company	1990-94	1995-99	2000-04	2005-07	1990-94	1995-99	2000-04	2005-07
	Million constant 2006 U.S. dollars				*Percent*			
By region:								
North America	1,956	2,191	2,481	2,471	0.82	1.18	1.23	1.74
Europe-ME	2,506	2,520	2,902	2,843	0.74	1.01	1.23	1.31
Asia-Pacific	2,945	3,272	3,205	2,826	1.25	1.82	1.93	2.57
Latin America	6	25	59	92	0.02	0.10	0.15	0.28
Global total	7,413	8,007	8,647	8,232	0.88	1.25	1.35	1.64
By country:								
U.S.	1,868	2,094	2,381	2,371	0.87	1.24	1.30	1.91
Canada	88	96	100	100	0.35	0.58	0.56	0.55
UK	467	413	470	397	0.89	1.08	1.26	1.26
Germany	286	300	311	293	0.40	0.59	0.72	0.81
France	450	450	555	618	0.87	1.23	1.74	2.18
Switzerland	318	333	336	319	2.90	4.97	5.75	6.29
Netherlands	266	316	295	236	1.54	2.36	2.35	1.93
Belgium	85	110	120	120	0.87	1.45	1.81	1.98
Italy	102	99	112	113	0.23	0.34	0.42	0.49
Spain	88	95	150	206	0.28	0.50	0.87	1.15
Sweden	93	49	48	48	1.39	0.95	1.05	1.25
Denmark	81	90	188	144	1.11	1.68	3.94	3.49
Norway	49	70	88	93	1.21	2.23	2.31	2.45
Finland	109	74	59	44	2.19	2.39	2.31	1.95
Czech Rep.	1	3	3	6	0.09	0.12	0.12	0.23
Hungary	1	4	5	6	0.03	0.25	0.21	0.32
Israel	7	7	10	4	0.30	0.29	0.44	0.22
Turkey	9	9	17	25				
South Africa			9	23				
Japan	2,609	2,857	2,666	1,990	1.34	1.94	1.97	2.40
South Korea	171	135	174	228	1.51	1.40	1.80	2.58
China, Taiwan		31	59	37		0.51	1.17	1.12
Australia	122	192	185	223	0.85	1.53	1.57	2.04
China			77	283				
Mexico	6	25	50	70	0.02	0.10	0.13	0.22
Chile			9	22				
Other	138	155	172	215				

Sources: USDA, Economic Research Service. R&D expenditures and GDP for the Food, Beverage and Tobacco manufacturing industry are from OECD (2010). Local currency nominal expenditures were converted to U.S. dollars using official exchange rates (World Bank) and then adjusted for inflation using the U.S. Gross Domestic Product implicit price deflator (*Economic Report of the President*, 2009).

Factors Affecting R&D Investment by Food Manufacturing Firms

Private firms may invest in R&D to develop new products or to reduce manufacturing costs (i.e., raise productivity of labor, capital, and materials). Such product and process innovations enable companies to maintain or expand market share, develop new markets, lower costs, and earn higher profits. While we have not found data that reveal how firms in the food manufacturing industry allocate R&D investments, we conjecture that most expenditures are likely allocated to new product innovation rather than cost-reducing processing innovations. Several factors influence decisions to engage in the various types of research in food manufacturing.

Consumer Demand

Consumer demand plays a large role in the nature of R&D in the food manufacturing industry. For example, innovations in poultry production often have been motivated by consumer demand for certain traits, such as greater amounts of white meat, pre-cooked products, and specialty shapes (e.g., "chicken nuggets"). Consumer convenience is a key factor, driving demand for time-saving products, such as bagged salad and prepared breakfast sandwiches (Martinez and Stewart, 2003). Additionally, consumers prefer variety in food products and product characteristics (Gopinath et al., 2003). According to Datamonitor, more than 20,000 new food products are introduced each year in the United States, although over 90 percent of these are classified as "not innovative" (USDA/ERS, 2010a). Examples of new product innovations include specialty Macaroni & Cheese from Kraft (e.g., the SpongeBob Squarepants variety) and private-label products (Martinez, 2007b; 2009). Delgado-Gutierrez and Bruhn (2008) note that high-valued characteristics of new food products include superior taste, longer shelf life, higher nutritional content, health benefits, and environmentally friendly packaging. These factors would motivate technical change that is focused on product development rather than process innovation.

New information or labeling requirements on the health attributes of food products can spur private R&D. Unnevehr and Jagmanaite (2008) found that new information on potentially adverse health effects of trans fats (and a new U.S. Food and Drug Administration regulation requiring disclosure of trans fat content on nutrition labels) led to fairly rapid development of new products low in trans fats. Innovations included the development of improved oilseed varieties, notably low linoleic soybeans, as well as dedicated supply chain coordination to produce trans fat-free food product alternatives.

Intellectual Property Protection

The role of intellectual property protection, such as patents and trademark protection, in motivating R&D spending by the food manufacturing industry is unclear. The industry is complex, with many heterogeneous products. Few new products succeed, and the industry is rife with product imitations (Gopinath and Vasavada, 1999). Patents are expensive to obtain, and new products with only minor modifications may not qualify for patent protection. Moreover, the effective market life of many new product innovations is often far less than the 20 years covered by patent protection. Thus, the transitory nature of food products may reduce the incentive to invest in such an expensive form of intellectual property protection (and the costs of enforcing it). Moreover, process innovations (innovations that improve manufacturing efficiency) are rarely patented (Gopinath et al., 2003). At the same time, intra-industry knowledge "spillovers" have been shown to play a significant role in the food processing industry (Gopinath and Vasavada, 1999). Spillovers are gains from research and innovation that benefit the industry as a whole but which cannot be fully captured by the firm(s) conducting the R&D. Significant R&D spillovers are likely to contribute to underinvestment by firms in R&D because innovators have difficulty earning the full returns to their innovations (Gopinath and Vasavada, 1999).

Industry Structure

The food manufacturing industry has an oligopolistic structure in which product markets tend to be dominated by a small number of large firms (Gopinath and Vasavada, 1999; Bolotova et al., 2007). Factors favoring this kind of market structure are economies of scale and scope found in processing, joint distributing, storing, and marketing (Bolotova et al., 2002). High sunk costs can act as a barrier to entry by new firms (Bolotova et al., 2007; Paul, 2000).

Industry consolidation (i.e., decreases in the number of firms) has been rising in a number of food manufacturing subsectors. Between 1972 and 1992, consolidation increased in eight U.S. food processing industries (Ollinger et al., 2005). These include meat (packing and processing), dairy (fluid milk and cheese processing), flour milling, corn milling, and feed and soybean processing.

Concentration (the relative size of the largest firms in an industry) has also increased as the structures of processing and manufacturing industries have changed. Concentration in beef packing has been a concern since the 1880s (Paul, 2001). Paul (2001) reports that the top four U.S. meatpacking companies accounted for 82 percent of the industry's output in 1994. In hog slaughter, the largest market share of the four firms reached 64 percent in 2004 (Martinez, 2007a). Concentration in the U.S. corn and flour-milling industries exceeded 70 percent as of 1992 (Ollinger et al., 2005). The industry has become increasingly vertically integrated (Henry and Rothwell, 1995; Martinez, 2002). Processed fruit and vegetable production is also often vertically integrated.

Using firm-level data for the U.S. food processing sector, Gopinath and Vasavada (1999) find positive correlations between patents and R&D and patents and market structure. Firms with higher market shares earned more patents, suggesting that they were also investing more in R&D. Additional evidence shows that concentration in food manufacturing is positively related to productivity of the sector, at least up to a point. Gopinath et al. (2003) find that concentration in food industries improved total factor productivity (TFP) growth in an invert-U fashion, with productivity growth initially rising with higher concentration but eventually slowing in industries that became too highly concentrated. Chan-Kang et al. (1999) find that productivity growth in the Canadian food processing sector fell behind that of the United States when U.S. firms engaged in extensive mergers. ChanKang et al. also find that R&D per unit of output for Canada was significantly less than that for the United States. They attribute Canada's general underinvestment in technical change to its failure to cut costs and merge manufacturers as the United States had done. Paul (2000) suggests that concentration and enhanced productivity may have been the result of the same stimuli, as indicated by cost economies, shortrun rigidities, innovation, and product differentiation.

Innovation and Productivity in Food Manufacturing

One reason to suspect a high share of R&D investment on product innovations and a low share on processing innovations is that growth in TFP in the food manufacturing industry is relatively low. The KLEMS-EU project has developed internationally comparable estimates of value-added TFP growth in primary and manufacturing industries, including the food manufacturing industry (O'Mahony and Timmer, 2009). Value-added TFP measures output (net of payments for energy and raw materials) relative to the capital and labor employed in the industry. At the industry level, growth in TFP primarily reflects process innovations that reduce labor and capital required to produce outputs.

Table 9.3. TFP growth in agriculture, food, and total manufacturing in OECD countries, 1980-2006

Country or region	Agriculture, forestry, and fishing	Food manufacturing	Total manufacturing
	TFP[1] index in 2006 with base year 1980=100		
U.S.	245.8	107.8	192.3
Eurozone[2]	288.1	104.2	149.2
UK	192.9	121.8	190.4
Japan	112.7	57.2	151.0

OECD = Organisation for Economic Co-operation and Development.

[1]Total factor productivity (TFP) is based on value-added output relative to capital and labor inputs employed in the industry.

[2]The Eurozone consists of Austria, Germany, France, Belgium, Netherlands, Luxembourg, Italy, Greece, Spain, Portugal, and Finland.

Source: USDA, Economic Research Service using EU-KLEMS.

From 1980 to 2006, TFP growth in food manufacturing was substantially below that in total manufacturing and agriculture in the United States, the UK, the "Eurozone,"[5] and Japan (table 9.3). While TFP in total manufacturing in the United States increased by 92 percent during the period, TFP in food manufacturing grew by only 7.8 percent. U.S. agricultural TFP growth grew by 146 percent over the same period. Other countries show similar patterns. Japanese food manufacturing actually registered a sharp fall in food manufacturing TFP even as its total manufacturing TFP grew by over 50 percent. The valued-added TFP indexes shown in the table indicate the rate of capital- and labor-saving technical change in an industry, and findings reveal that relatively little of this innovation occurred in food manufacturing overall.

Gopinath and Roe (1996) suggest that many of the sources of productivity growth in food processing lie outside the sector, especially through linkages with primary agriculture. Using U.S. data from 1960-91, they find that the food manufacturing industry benefitted from productivity growth in the primary agricultural sector, which led to more abundant, lower cost raw materials for processing. Paul (2000) cites studies that found that productivity growth in the agricultural production sector (i.e., before the farm gate) reduced costs in the food processing industry in the United States and the UK.

Conclusions

Firm-level data on food manufacturing R&D are limited and incomplete. Data from OECD's Business Expenditure on R&D, which provides industry-level data by country, suggest that expenditures on food manufacturing R&D increased, both in real expenditures and as percentage of industry value added. Research intensity in food manufacturing, however, is low compared with that in manufacturing as a whole and in other agricultural input industries.

Among the drivers of R&D in this sector, consumer demand is particularly important. Convenience is one of the most desired traits among consumers; others are quality and value. Consumers expect many new products with these characteristics; as a result, the food manufacturing system moves swiftly in the area of technical change. Most of the R&D in the industry appears to be directed to product, as opposed to process, innovations.

Formal intellectual property protection plays a smaller role in motivating research. The industry abounds with imitators, and products change quickly. While patent protection is used, it is generally too costly for most innovations, particularly because the 20-year period of protection provided by patents usually is not needed.

The food manufacturing industry is oligopolistic in structure. The level of industry concentration in subsectors of the food industry may affect incentives for research. Studies have found that concentration is correlated with productivity growth in the industry. Thus, concentration is most likely positively related to research investments. Productivity growth in food manufacturing (as measured by value-added TFP), however, has been small relative to that of total manufacturing and agriculture. The value of some kinds of innovations, such as new products with novel characteristics, may not be reflected in this TFP measure.

References

Alston, J., M. Anderson, J. James, and P. Pardey. 2010. Persistence Pays: U.S. Agricultural Productivity Growth and the Benefits from Public R&D Spending, New York: Springer.

Bolotova, Yuliya, John M. Connor, and Douglas J. Miller. 2007. "Factors Influencing the Magnitude of Cartel Overcharges: An Empirical Analysis of Food-Industry Cartels," Agribusiness 23 (1): 17-33.

Chan-Kang, Connie, Steven Buccola, and Joe Kerkvliet. 1999. "Investment and Productivity in Canadian and U.S. Food Manufacturing," Canadian Journal of Agricultural Economics 47: 105-118.

Delgado-Gutierrez, C., and C. M. Bruhn. 2008. "Health Professionals' Attitudes and Educational Needs Regarding New Food Processing Technologies," Journal of Food Science Education, Vol. 7, pp. 78-83.

EU KLEMS. 2011. "Growth and Productivity Accounts: November 2009 Release," updated March 2011, available at: www.euklems.net/ (accessed May 5, 2011).

Gopinath, Munisamy, Daniel Pick, and Younghai Li. 2003. "Concentration and Innovation in the U.S. Food Industries," Journal of Agricultural and Food Industrial Industries, 1 (1) article 15, pp. 1-21.

Gopinath, Munisamy, and Terry L. Roe. 1996. "Sources of Growth in U.S. GDP and the Economy-Wide Linkages to the Agricultural Sector," Journal of Agricultural and Resource Economics 21 (2): 325-340.

Gopinath, Munisamy, and Utpal Vasavada. 1999. "Patents, R&D, and Market Structure in the U.S. Food Processing Industry," Journal of Agricultural and Resource Economics 24: 127-139.

Henry, Richard, and Graeme Rothwell. 1995. The World Poultry Industry, The World Bank.

Huang, Kuo S. 2003. Food Manufacturing Productivity and Its Economic Implications, Technical Bulletin No. 1905, U.S. Department of Agriculture, Economic Research Service, available at: www.ers.usda.gov/ publications/tb1905/

James, Clive. 1997. Progress in Public-Private Sector Partnerships in International Agricultural Research and Development, ISAAA Briefs No. 4-1997, International Service for the Acquisition of Agri-Biotech Applications.

Klotz, C., K. Fuglie, and C. Pray. 1995. Private Sector Agricultural Research Expenditures in the United States, 1960-1992, Staff Paper AGES-9525, U.S. Department of Agriculture, Economic Research Service.

Martinez, Steve. 2010. "Recessions Brings Record Number of New Store-Brand Offerings," Amber Waves, Vol. 8, Issue 2, U.S. Department of Agriculture, Economic Research Service, available at: www.ers.usda.gov/ amberwaves/june10/findings/ newstorebrand. htm.

Martinez, Steve. 2007(b). "Food Product Introductions Continue to Set Records," Amber Waves, Vol. 5, Issue 5, U.S. Department of Agriculture, Economic Research Service, November, available at: www.ers.usda.gov/amberwaves/november07/findings/ food product.htm

Martinez, Steve. 2007(a). The U.S. Food Marketing System: Recent Developments, 1997-2006, Economic Research Report No. 42. U.S. Department of Agriculture, Economic Research Service, May, available at: www.ers.usda.gov/publications/err42/

Martinez, Steve. 2002. Vertical Coordination of Marketing Systems: Lessons From the Poultry, Egg and Pork Industries, Agricultural Economic Report No. 807, U.S. Department of Agriculture, Economic Research Service, May, available at: www.ers.usda.gov/publications/aer807/

Martinez, Stephan and Hayden Stewart. 2003. "From Supply Push to Demand Pull: Agribusiness Strategies for Today's Consumers, Amber Waves, Vol. 1, Issue 5, U.S. Department of Agriculture, Economic Research Service, November, available at: www.ers.usda.gov/amberwaves/november03/ findings/pushdemandpull.htm

Ollinger, Michael, Sang V. Nguyen, Donald Blayney, Bill Chambers, and Ken Nelson. 2005. Structural Change in the Meat, Poultry, Dairy, and Grain Processing Industries, Economic Research Report No. 3, U.S. Department of Agriculture, Economic Research Service, October, available at: www. ers.usda.gov/publications/err3/

O'Mahony, Mary, and Marcel P. Timmer. 2009. "Output, Input and Productivity Measures at the Industry Level: The EU KLEMS Database," Economic Journal 119 (538): F374-F403.

Organisation for Economic Co-operation and Development (OECD). "Structural Analysis Statistics (STAN) Database," OECD, Paris, available at: www.sourceoecd.org/database/ OECDStat (accessed June 27, 2011).

Pardey, Philip G., Nienke Beintema, Steven Dehmer, and Stanley Wood. 2006. Agricultural Research: A Growing Global Divide? Food Policy Report, International Food Policy Research Institute.

Paul, Catherine Morrison. 2001. "Market and Cost Structure in the U.S. Beef Packing Industry: A Plant Level Analysis," American Journal of Agricultural Economics 83(1): 65-76.

Paul, Catherine Morrison. 2000. "Modeling and Measuring Productivity in the Agri-Food Sector: Trends, Causes and Effects," Canadian Journal of Agricultural Economics 48: 217-240.

Unnevehr, Laurian J., and Evelina Jagmanaite. 2008. "Getting Rid of Trans Fats in the U.S. Diet: Policies, Incentives and Progress," Food Policy 33: 497-501.

U.S. Department of Agriculture, Economic Research Service (USDA/ERS). 2010a. "Food Marketing System in the U.S.: New Product Introductions," available at: www.ers.usda.gov/briefing/foodmarketingsystem/new_ product.htm (accessed February 6, 2011).

U.S. Department of Agriculture, Economic Research Service (USDA/ERS). 2010b. "Agricultural Research Funding in the Public and Private Sectors," available at www.ers.usda.gov/data/agresearchfunding/ (accessed May 4, 2011).

U.S. Department of Commerce, U.S. Bureau of the Census. 2010. 2009 Annual Survey of Manufactures, available at: www.census.gov/ manufacturing/asm/

World Bank. World Development Indicators Databank. World Bank, Washington, DC. Available at http://data.worldbank.org/data-catalog/ world-development-indicators

PRIVATE RESEARCH AND DEVELOPMENT FOR BIOFUEL

Carl E. Pray, Rupa Karmarkar-Deshmukh and Keith O. Fuglie

Agriculture in the 21st century faces two major challenges: meeting the rising food and livestock feed demand from a growing and wealthier world population and helping to meet the rising demand for biofuels for transportation. The second and "newer" challenge is driven by high oil prices, government policies, and increased societal concerns about the role of fossil fuels in global warming.

Until a decade ago, agriculture had met such challenges by increasing the amount of land area under cultivation globally and increasing crop and livestock yields through the development and application of new farming practices and technologies.

Since 2006, the simultaneous expansion of biofuel production and a rapid rise in food prices has raised concerns that agricultural supply might not be keeping pace with these increased demands. The term "biofuel" encompasses all types of renewable fuels that are derived from biological feedstocks and are used as transportation fuel, as fuel for domestic cooking and heating, or as an energy source for industry.

Ethanol (which is also used as an industrial solvent and in alcoholic beverages) is the most popular biofuel and is made from crops with a high starch or sucrose content, such as corn and sugarcane. Biodiesel, the second most widely used biofuel, is extracted from oil-bearing seeds, such as soybean, rapeseed (canola), and crude palm oil.

We find that private-sector investment in biofuel research was about $1.47 billion worldwide in 2009. In comparison, the major private oil companies in high-income countries and Brazil spent at least $6 billion on R&D in the same year—primarily on R&D on fossil fuels.[1]

The biofuel research investment is a relatively small amount compared with the $10.4 billion spent by the private sector on agricultural input research in 2009 and the $11.5 billion spent on food industry research in 2007 (the last year of comparable data - see chapter 1).

Global Production of Biofuels

A combination of such factors as climate change, energy security, and oil prices has led to the biofuel revolution, which has increased the demand for ethanol and biodiesel. Global biofuel production has more than doubled between 2005 and 2009 (fig. 10.1).

Government policies in the United States; the EU; and some developing countries, such as Brazil, India, and China, require that a certain percentage (5-25 percent) of automotive fuel used consist of ethanol or biodiesel. These policies, along with increasing public support for a reduction in greenhouse gas emissions from fossil fuels, have contributed to the sharp increase in demand for biofuels. Supply of biofuels has often been subsidized at either the biofuel factory level or the feedstock level. The combined effect of these policies has driven much of the expansion of biofuel production.

Among all countries, the United States and Brazil have dominated ethanol production, and Germany and the United States have led in biodiesel production (table 10.1). These countries, which have the largest domestic market demand, have been in the forefront of biofuel production. Brazil pioneered commercial biofuel production due to a combination of optimal growing conditions for sugarcane (the most efficient biofuel feedstock) and early government intervention, such as mandates, production subsidies, and research and development programs to increase biofuel productivity and develop engines that could run on biofuel. These initiatives led to rapid increases in ethanol production and the manufacture and rapid spread of vehicles that run on 100 percent ethanol, starting in 1979 (Matsuoka et al., 2009). Some developing countries, such as India and China, are also beginning to meet some of their energy and fuel demand through policies and initiatives to support production and use of energy crops.

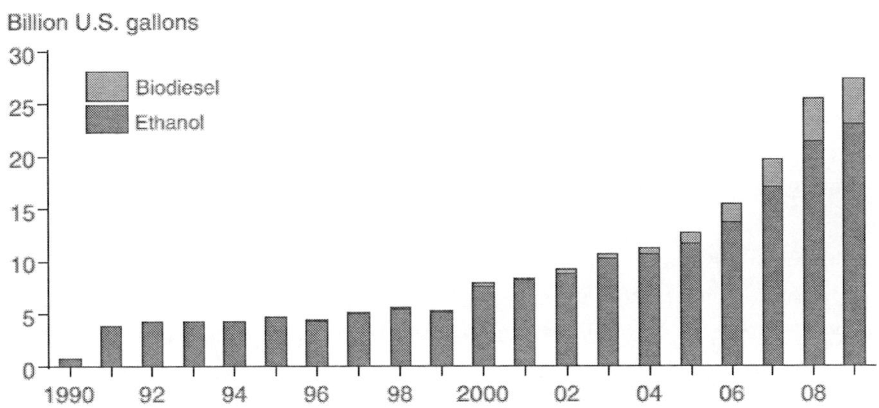

Source: USDA, Economic Research Service using F.O. Licht (2009).

Figure 10.1. Global biofuel production Billion U.S. gallons.

As the governments of more countries require the use of biofuels, they may increasingly rely on trade to meet domestic biofuel demand. Developing countries in tropical regions often have a competitive advantage in biofuel production. Brazil has become the largest exporter of biofuel, posting a record 1.365 billion gallons of ethanol shipped in 2008, an increase of 45 percent over 2007 (*Biofuels Digest*, 2009a). Brazil's rapid growth in biofuel production can be attributed to land available for cropland expansion and the high (and rising) productivity of its sugarcane-growing system. These factors have also enabled Brazil to expand food production, as land planted to food crops has kept pace with the area sown to sugarcane (F.O. Licht, 2008).

Table 10.1. Countries that produced the most biofuel in 2008

Country	Share of global biofuel production	Main feedstock
	Percent	
Ethanol		
United States	51	Corn
Brazil	38	Sugarcane
European Union	4	Sugar beets
China	3	Corn
India	0.6	Sugarcane
Global ethanol share for countries listed	96.6	
Biodiesel		
Germany	19	Rapeseed
United States	17.5	Soybean
France	14	Rapeseed
Brazil	10	Soybean, castor
Argentina	5	Soybean
Thailand	3.5	Palm oil
Global biodiesel share for countries listed	69.0	

Source: USDA, Economic Research Service using data from F.O. Licht (2008).

The structure of the biofuel industry is illustrated in figure 10.2. Agricultural input firms sell seeds, fertilizer, farm machinery, and other inputs to farmers. Farmers produce feedstock, such as corn, sugarcane, or rapeseed, and sell it to ethanol and biodiesel producers. Biofuel producers combine the feedstock with enzymes, yeasts, thermal energy, labor, and capital to make the ethanol and biodiesel. Producers usually sell the ethanol and biodiesel to blenders, who add the product to gasoline or diesel produced from fossil fuel. The blenders then sell the blended fuel to consumers for their autos, tractors, trucks, and, in the future, possibly airplanes.

In 2011, virtually all commercial biofuel is from first-generation feedstocks (i.e., sugarcane, corn, soybean, and rapeseed). Ethanol and biodiesel are produced by major agribusinesses, chemical companies with agricultural or manufacturing infrastructure in place, or specialized biofuel companies. Among the big commodity companies, ADM is leading in biofuel production, with Cargill and Bunge making much smaller investments. POET leads the independent biofuel producers and is the leading producer of ethanol for biofuel in the United States. The Spanish company Abengoa, which is involved in telecommunications and transportation as well as alternative energy, has invested substantially in biofuels in the United States as well. Oil companies have also entered into the business—Valero Energy Corporation, the largest U.S. oil refinery company, bought the biofuel company VeraSun Energy in 2009 after VeraSun went bankrupt. Shell has entered into a joint venture with Cosan, the largest sugarcane-based ethanol producer in Brazil, and Petrobras has been buying sugar and ethanol companies in Brazil. But so far, the oil companies are relatively small players in the production of first-generation biofuels.

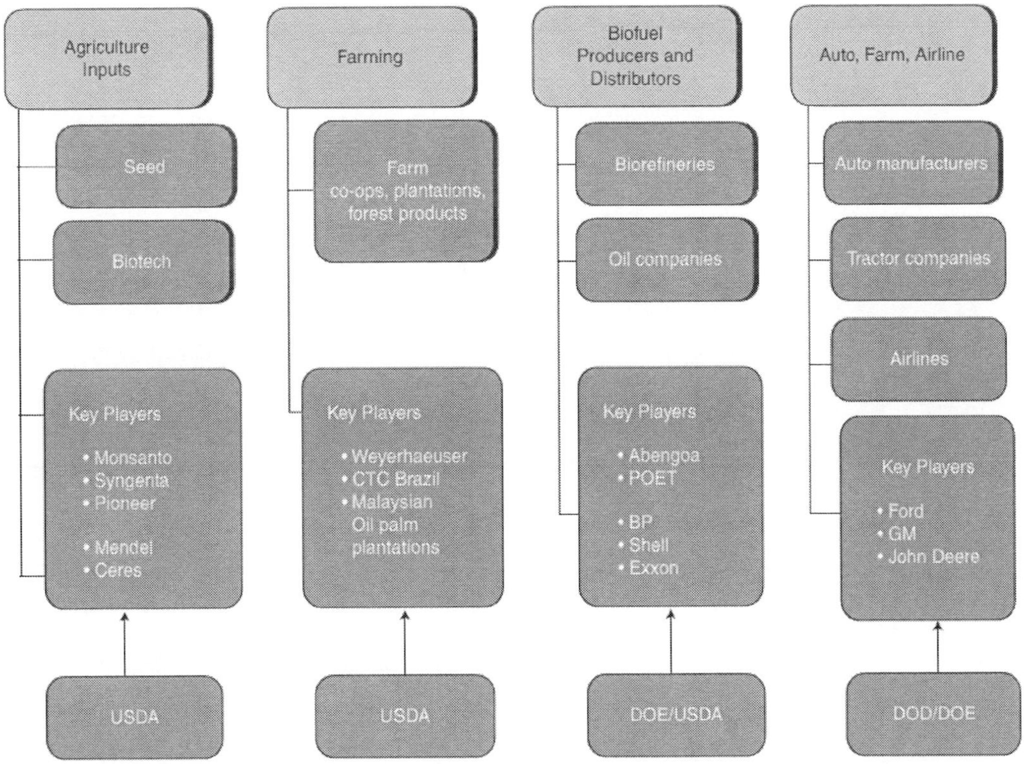

The last row indicates major Federal agencies that are funding biofuel research in the private sector:
USDA = U.S. Department of Agriculture.
DOE = U.S. Department of Energy.
DOD = U.S. Department of Defense.
Source: USDA, Economic Research Service.

Figure 10.2. The U.S. biofuel industry supply chain.

Biofuel R&D Conducted by Industry

It is relatively easy to find general information about the kind of research and technology being conducted by development firms, but it is considerably more difficult to determine the amount of money being spent on biofuel R&D by companies in the biofuel supply chain. Most companies that produce only biofuel are privately held and do not report their R&D. For most other companies in the supply chain, biofuels or the inputs for biofuel production account for a small part of their business and their R&D expenditures. Although many of these companies are publicly held and report their total R&D spending, almost none reveal R&D allocations toward improving biofuel production or use. The estimates that we provide are based on the scattered data available from interviews with company officials, company websites and annual reports, or the press. Using this information, we estimate expenditures on biofuel-related R&D for each segment of the supply chain globally in 2009.

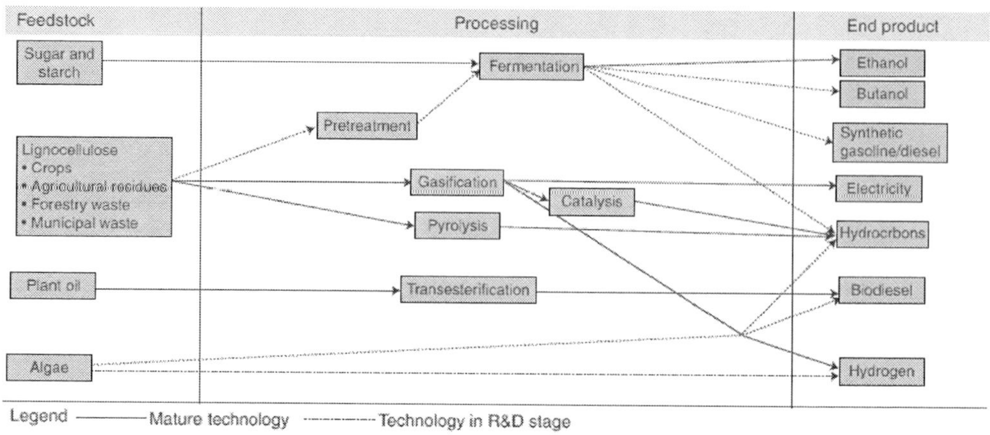

Source: USDA, Economic Research Service using Rajagopal et al. (2009).

Figure 10.3. Current and potential pathways to bioenergy production.

Companies and government research institutes use R&D to develop a variety of pathways to bioenergy production (fig. 10.3). Companies are working on improving a variety of feedstocks ranging from corn and sugarcane to algae. Improvements in processing are also being explored to make biofuel economically viable and environmentally friendly. Finally, companies are continually conducting research on the many different forms of bioenergy. Some of these end products, such as ethanol, have to be blended with gasoline to fuel most current vehicles; others are hydrocarbons that can be directly substituted for gasoline or diesel—these are now called "drop-in" biofuels.

Some firms specialize in improving or reducing the cost of one component of one pathway. For example, Mendel Biotech is specializing in improving the productivity of cellulosic feedstuffs, such as miscanthus. Other firms are working on an entire pathway or several pathways, usually in collaboration with other firms. As these pathways get closer to commercialization, the research focus will shift to increasing the efficiency of the entire supply chain (Richard, 2010). Collaborations, joint ventures, mergers, and acquisitions have been driven by the importance of increasing the efficiency of the entire supply chain as new forms of biofuel become commercially viable.

The first component of the biofuel supply chain contributing to biofuel research is agricultural input companies that work to improve the productivity of biofuel feedstocks, such as crops, grasses, wild plants, and algae. This group includes large companies, such as Monsanto, Syngenta, DuPont, and BASF, as well as some small biotechnology firms. These firms are developing new crops, new varieties of existing crops, and new traits for these crops that will maximize their usefulness as biofuel feedstocks.

Major seed-biotechnology-chemical input developers, such as Monsanto, Syngenta, and DuPont, recognize the profitability of the biofuel crop market for corn (their major seed crop) and have shifted R&D resources to screen elite corn lines and develop hybrids that can produce more ethanol per acre. They also seek to develop varieties of new energy crops (e.g., switchgrass, miscanthus, and jatropha) through inhouse research and investments in small biotech companies. For example, Monsanto is an equity shareholder of both Mendel Biotech and Ceres, Inc. In addition, both Syngenta and Monsanto have made major investments in sugarcane R&D. In 2008, Monsanto spent $290 million to buy two Brazilian companies,

CanaVialis and Alellyx, which were considered leaders in sugarcane breeding and biotechnology (Monsanto, 2008). Together, these firms spent $32 million on sugarcane R&D in 2008 (BNDES & CGEE, 2008, p. 65). Syngenta has been collaborating with the Queensland Institute of Technology (Australia), the Agronomy Institute of Campinas in Sao Paulo, Brazil, and several other organizations to develop technologies for cellulosic ethanol from sugarcane (Syngenta, 2007). In 2009, BASF also made a commitment to work on sugarcane in Brazil (BASF, 2009).

The major agricultural seed-chemical-biotechnology companies collectively spent $2.08 billion per year on average for crop breeding and biotechnology research in 2007-09 (see table 2.3). We estimate that about 10 percent of that amount, or $208 million per year, was on biofuel-related R&D. We then add the $32 million that was being spent by Monsanto's new Brazilian acquisitions on sugarcane/biofuel R&D[2] for a total of $240 million.

Several small biotechnology firms have been conducting research to develop feedstocks with higher cellulosic content and lower lignin content for more efficient net energy output (Sticklen, 2007). For example, Ceres, Inc. has been investigating higher yielding switchgrass; Edenspace Corporation has been developing commercial corn hybrids that have lower cost conversion to ethanol (BRDI, 2007); Mendel Biotechnology, Inc. has been developing miscanthus varieties that are high yielding, can be propagated commercially by seed, and are not invasive. Industry sources estimate that the small firms focusing on developing cellulosic biomass for bioenergy, led by Ceres, Mendel Biotechnology, and the forestry breeding and biotechnology firm Arborgen, spent at least $50 million on biofuel-related R&D in 2009.

The next component of the biofuel supply chain is the farmer who produces the corn, sugarcane, soybeans, palm oil, and biomass and the forestry industry that produces woody biomass. Few farms in the United States or the EU are large enough to conduct their own R&D, but a number of farmers have established cooperatives to invest in ethanol production from corn. A few of these biofuel cooperatives may be supporting research and engineering activity to improve the efficiency of their biofuel production. In Brazil, sugarcane producers and sugar mill owners who also run sugarcane plantations tax themselves to support $23 million of sugarcane and ethanol research at Centro de Tecnologia Canavieira (CTC) (BNDES & CGEE, 2008, p. 164). Some of the largest plantation companies in Malaysia and Indonesia are big enough to capture benefits from research and conduct substantial research on biofuel from crude palm oil. Three major Malaysian plantation companies—Sime Darby, Asiatic Development Berhad, and IOI-Golden Hope—spent over $40 million on R&D in 2008 according to their annual reports. Most of this work focused on improving palm oil production, but some of it is specifically for biofuel technology from palm oil and for second-generation biofuels[3] using waste from palm oil processing and palm oil trees. In addition, these companies are collaborating with biotech companies, such as U.S.-based Synthetic Genomics, to map the palm oil and jatropha genomes (ACGT, 2010).

Some forestry companies have announced investments in biofuels R&D. For example, in 2008, Weyerhaeuser and Chevron announced a new joint venture called Catchlight Energy, which was aimed at developing biofuels from forest-based material. In 2009, Weyerhaeuser spent $51 million on research, including $22 million in its "corporate and other" business segment where it records financial information from its Catchlight Energy joint venture (Weyerhaeuser, 2009a; 2009b). It is reasonable to believe that most of this $22 million "corporate" R&D is directed toward developing biofuels from forestry biomass. Most of the research conducted at CTC is also biofuel related. In contrast, most of the palm oil companies' R&D is not spent on biofuels. R&D expenditures allocated to biofuels by this group may account for only about a quarter of total R&D investments, or about $10 million per year. Until we can obtain more accurate numbers, we assume that the Southeast Asian plantation companies, CTC, and the forestry companies spent at least $50 million on biofuel research in 2009.

Companies that seek to use algae as a feedstock for biodiesel are also making substantial investments in biofuels-related R&D. Some of the largest companies in the United States engaged in algal research for biofuel are Algenol, Sapphire, Solayzmes, and Synthetic Genomics. Scientists associated with the industry reported that algal fuel companies are spending about $200 million annually on biofuels. Of that amount, Synthetic Genomics is reportedly receiving $60 million a year from ExxonMobil (Chang, 2009).

Among the current U.S. biofuel producers, a number of firms are conducting R&D activities, many with the support of the U.S. Government. ADM has a joint venture with Conoco Phillips to develop second-generation biofuels. ADM's total research budget was $45 million in 2008 (ADM, 2009). POET is investing in research to improve the efficiency of its corn-based ethanol production. In addition, POET is working with Novozymes and others to develop a cellulosic ethanol facility. The U.S. Department of Energy (DOE) is providing up to $80 million to help fund this effort. Under the terms of the grant, POET is required to match the DOE funding (POET 60 percent, DOE 40 percent) over 2 years. Industry representatives report that most of the research by medium-sized and smaller biofuel companies is financed by DOE or USDA. The smaller companies concentrate primarily on second-generation biofuels, while some of the larger companies (such as POET and ADM) conduct substantial, inhouse research on first- and second-generation biofuels.

R&D expenditure data for biofuel firms (other than oil companies) outside the U.S. are equally difficult to find. One European biofuel firm for which data on R&D are available is Neste Oil. Neste is developing processes that can produce biodiesel from all types of vegetable oils and fats and spent $52 million on R&D in 2009 (Neste Oil, 2010). We estimate that biofuel producers spent about $100 million for R&D in 2009: $40 million to $50 million by U.S. biofuel producers plus the $52 million reported by Neste Oil.

In recent years, a few large oil and gasoline companies have also started to invest in biofuel R&D. So far, these companies are focusing investments on first-generation biofuels—mostly in Brazilian sugarcane. For example, in 2008, BP announced an agreement to invest $1 billion in a Brazilian sugarcane-based ethanol factory (BP, 2008), and Shell recently agreed to a joint venture with Cosan in Brazil. R&D investments of these companies are also targeted at the development of second-generation biofuels from cellulosic sources—sugarcane bagasse, corn stover, miscanthus, switchgrass, and wood waste—and third-generation biofuel feedstocks such as algae. The oil and gas companies are also conducting or financing research and engineering development on the transportation and blending of ethanol and the effects of different ethanol blends on engine performance.

Shell reports that it has been engaged in biomass R&D for 30 years and has biofuels research and technology centers in Chester, UK; Houston, TX (U.S.); Amsterdam, Netherlands; and Bangalore, India (Royal Dutch Shell, 2008). In addition, Shell collaborates on biofuels R&D with six university and government research programs in the United States, the UK, China, and Brazil and has equity stakes in the small cellulosic biofuels research companies Codexis and Iogen. BP is investing $50 million a year for 10 years in biofuel R&D at the Energy Biosciences Institute (EBI), which has its headquarters at the University of

California, Berkeley. EBI is a partnership between BP, UC-Berkeley, DOE's Lawrence Livermore Laboratory, and the University of Illinois. BP also has its own inhouse biofuels research program and research collaborations with other companies and institutes around the world.

The only oil and gas companies that have stated the amount of their investment in biofuel R&D are ExxonMobil and Petrobras. ExxonMobil announced in 2009 that it planned to invest $600 million over 5 years in research on biofuels from algae (Howell, 2009). Half of this money would be spent on research by the biotech company Synthetic Genomics and the other half would be spent on inhouse research. Petrobras is spending about 8 percent of its total R&D budget on biofuel, or $80 million annually (Petrobras, 2010). ExxonMobil's annual expenditure of $120 million ($600 million prorated over 5 years) is about 14 percent of its total corporate research budget in 2008. If we assume that ExxonMobil and other major oil firms (BP, Shell, Chevron, Total, ENI, and ConocoPhillips) are also investing that percentage of their total research in biofuels R&D and that Petrobras is spending $80 million, then investment by this segment of the biofuels supply chain would be $677 million in 2009. This is likely an underestimate of the global total by oil companies because many national oil companies like Malaysian-based Petronas, Chinese National Petroleum Corporation, China National Offshore Oil Corporation, and Indian National Oil Company are also investing in research on biofuel.

The newest startup biofuel companies are focusing their research on developing processes that will convert sucrose or biomass into "drop-in biofuel," also known as green gasoline or green diesel (in Brazil it is called Canadiesel because it is made from sugarcane). These companies use engineered yeasts and bacteria to produce biofuels that are identical in chemical makeup to gasoline or diesel and therefore do not require blending with fossil fuels. Several leaders in this field—Amyris, Codexis, and Gevo—went public and reported a total of $82 million in R&D spending for 2009.

Enzymes and yeasts are important inputs to the bioethanol conversion process because the complex carbohydrates in biomass need to be broken down into simpler sugars that can then be fermented to produce ethanol (Sheehan and Himmel, 1999). The critical difference between feedstock conversion technologies is the enzyme or bacteria used to treat the biomass (because using the right enzymes for a given feedstock is crucial to obtaining energy-efficient ethanol at the end of the process). Novozymes is the biggest supplier of enzymes for first-generation biofuels, followed by Genencor. Diversa, an enzyme company that became part of Verenium in 2007, is another major player and is partnered in a joint venture with BP to produce second-generation ethanol from sugarcane bagasse in Florida. Novozymes and Verenium reported total R&D investments of $60 million to $70 million in 2009. Based on interviews with industry members, we estimate that these firms spent about a sixth of their total R&D on biofuels (with Genencor spending a similar amount), for a total R&D investment of $31 million annually by the enzymes supply component of the biofuels supply chain. The other suppliers of inputs to the biofuel industry are firms that build equipment for the biofuel factories. Some of the leaders in this industry are ICM and UOP (Honeywell) in the United States, Dedini in Brazil, and Praj in India and elsewhere. Praj reported investments in R&D of $4 million in 2009-10 million annually (Praj, 2010-2011), and Dedini contributed about $10 million to the Sao Paulo government's biofuel research program. We do not have data for the other companies that build factory equipment but assume that they invest amounts about equal to that of Dedini. Total investments for the four companies are thus estimated at

$40 million per year. At the end of the supply chain are consumers who fill up their vehicles with the blended gasoline or biodiesel or other form of biofuel. Car manufacturers such as Ford and General Motors, tractor manufacturers such as John Deere and Mahindra & Mahindra, and the companies that supply parts to these companies have shifted some of their research and engineering efforts to develop modified engines that accommodate the properties of the new biofuels or biofuel blends. The manufacturers of airplane engines and airline companies themselves have been testing and modifying their engines to use biodiesel. However, we do not have any information on the investment amounts that these companies may be making in biofuel R&D and therefore exclude this segment of the supply chain from our estimates. For 2009, we estimate total private-sector investment in biofuels R&D at $1.47 billion (table 10.2). Although we do not have detailed data on trends in biofuel R&D, it is clear that investments have risen in recent years, particularly as oil companies have started investing in research on second- and third-generation biofuels. Shell has the oldest continuous biofuel program, but between 2007 and mid-2009, it reportedly quadrupled its biofuel research spending (Krauss, 2009). Most of the other large oil companies appear to have started their biofuel R&D programs after 2000. BP's big investment in the Energy Bioscience Institute started in 2007. ExxonMobil was conducting some biofuel research before 2009, but the investment was a minor one compared with the $120 million per year program it announced that year. Research by the smaller biofuel companies also appears to have accelerated since 2000. Before then, investments in research on grasses for biomass or algal-based biofuels were scant. Research by small companies on conversion of sugars directly into green fuels is also new since 2000.

Table 10.2. Biofuel research and development (R&D) by private firms in 2009

Type of firm	R&D
	Million U.S. dollars
Large agricultural seed-biotechnology-chemical input companies	240
Small biotechnology companies developing cellulosic biomass	50
Plantations, farms, and forest product companies	50
Algal biofuel companies	200
Biofuel producers	100
Oil companies	677
Producers of green gas and green diesel	82
Enzyme companies	31
Biofuel equipment manufacturers	40
Car, tractor, airlines	n.a.
Total	1,470

n.a. = not available or unknown.

Sources: USDA, Economic Research Service using corporate annual reports, press reports, and author interviews with industry representatives.

Public Funding for Bioenergy R&D

Government support for bioenergy R&D began in the 1970s following the energy crises of that period. Funding of bioenergy R&D in OECD countries[4] waned in the 1980s but began to recover in the 1990s and then tripled between 2000 and 2007, rising from around $200 million per year to over $600 million per year (fig. 10.4). R&D bioenergy expenditures fell after 2007 in many OECD countries with the onset of the global financial crisis and economic recession but were boosted in 2009 in the United States by economic stimulus funding provided by the American Recovery and Reconstruction Act of 2009 (ARRA).

Historically, the U.S. Government has invested more on biofuel-related R&D than any developed country, accounting for nearly 40 percent of total public bioenergy R&D by OECD countries since 1974 (IEA, 2010). Other countries with major bioenergy R&D programs include Canada, Japan, and Sweden. Outside of the OECD countries, major investments in biofuel R&D have also been made by Governments of Brazil, China, and India, but public-sector R&D data from those countries are not available.

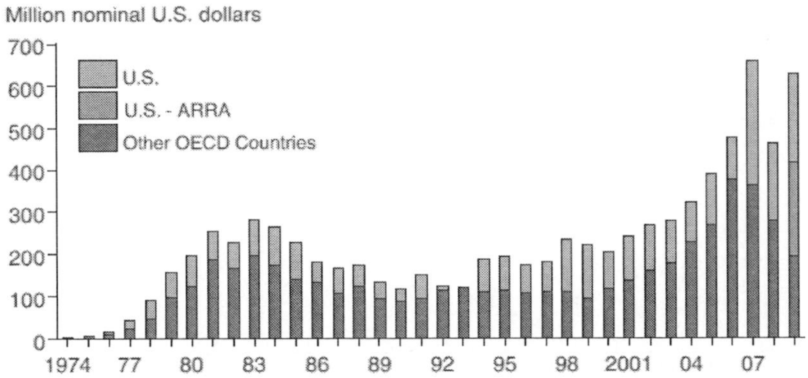

Note: USA-ARRA is a one-time increase in spending due to the American Recovery and Reconstruction Act (economic stimulus funding). OECD is the Organisation for Economic Co-operation and Development; membership comprises most high-income nations.

Source: USDA, Economic Research Service using International Energy Administration.

Figure 10.4. Government funding of bioenergy research and development (R&D) in OECD countries.

Government funding for bioenergy R&D may be provided to and carried out at government laboratories, universities, and private companies. One question regarding private bioenergy R&D is the relative importance of the government as a source of scientific discoveries and innovations as well as a source of funds. The rapid growth in private bioenergy R&D since 2000 is in part due to new commercial opportunities created by past public investments in bioenergy science and technology. Growth may also simply be due to a rise in government subsidies provided to private R&D. We examined in more detail the role and significance of government funding of private bioenergy R&D in the United States, which has world's largest bioenergy R&D program. For over two decades, DOE had been the primary government funder of biofuel research since interest in this technology arose in the 1970s. DOE mainly supported R&D on processing and conversion technologies rather than improving feedstock production. In 2000, DOE and USDA formed the Joint Biomass R&D Initiative (BRDI) to provide financial incentives to public and private institutions to undertake R&D for biofuels. BRDI, under which each agency takes individual responsibility for projects, mainly focuses on "plant science research" and "biorefinery demonstration and deployment" types of projects, as well as feasibility studies on next-generation technologies, such as synfuels. The total funds allocated by BRDI from 2002 to 2006 were about $160 million (BRDI, 2007). DOE's share was $130 million, of which 71 percent went to biofuel producers. USDA contributed $30 million to BRDI, with 41 percent going to ethanol producers. In many cases, private firms are required to match a portion of the government grant. For example, in 2007, BRDI solicitations required firms to match 20 percent of the grant money for research projects and up to 50 percent for demonstration projects. More recently, in January 2009, BRDI announced awards up to $25 million for R&D on technologies and processes to produce biofuels, bioenergy, and high-value biobased products. In 2009, DOE spent about $200 million on biofuel research and USDA spent about $69 million (table 10.3). DOE started funding three Bioenergy Research Centers in 2008 (with 2009 funding of about $75 million) (see table). These centers, which are headquartered in California, Wisconsin, and Tennessee, bring together researchers from 18 universities, 7 DOE national laboratories, at least 1 nonprofit organization, and many private companies. DOE also funds research on algal biofuel at its National Renewable Energy Laboratory in Colorado, and new research programs are scheduled to start under funding through the ARRA. USDA's ARS invests about $31 million annually in biofuel R&D. USDA also distributes $20 million per year to land grant universities and other institutions for biofuel research through the National Institute of Food and Agriculture (NIFA) and $18 million to BRDI.[5]

Table 10.3. Funding of bioenergy research by the U.S. Departments of Energy and Agriculture in 2009

Company	U.S. Department of Energy (DOE)	U.S. Department of Agriculture (USDA)[1]	Joint DOE-USDA biomass R&D initiative (BRDI)
Total biofuel research	Approximately $200 million	$69 million	$24.4 million ($4.9 million DOE, $19.5 million USDA)
Inhouse	n.a.	$31 million by Agricultural Research Service (ARS)	None
	$75 million to three bioenergy research centers (BRCs)		
	$5 million per year for 5 years from ARRA1 for BRCs	$20 million by National Institute for Food & Agriculture (NIFA) to Land Grant and other universities	
Extramural	$39.4 million by Advanced Research Projects Agency-Energy for biofuel research		$24 million to companies and universities in competitive grants
	$85 million from ARRA2 over 2010-13 for two new biofuel consortia		

n.a. = not available.

[1]This is a conservative estimate of USDA's biofuel-related research. The figure of $31 million for ARS biofuels research relates specifically to the Bioenergy National Program and does not include research on genetic development of key crops. In addition, the USDA total reported in the table does not include biofuels research by USDA's Forest Service.

[2]ARRA = American Recovery and Reconstruction Act of 2009.

Source: USDA, Economic Research Service author interviews and DOE (2009).

While BRDI grants have clearly been important to a number of firms, total government funding of private biofuel R&D is only a small share of private funding of its own inhouse R&D (see table 10.2). However, the Federal Government appears to be the primary source of funding of biofuel R&D at U.S. universities. While some private companies like BP (see previous section) have made major resource commitments to universities, these appear to be far smaller than the extramural biofuel R&D funding provided by DOE and USDA (estimated at $225 million in 2009 in table 10.3).

The Future of Biofuel Research and Biofuels Technology

The major investments in biofuels research by both the public and private sector appear to be directed toward reducing the cost of producing second-generation biofuels from biomass such as corn stover, bagasse, miscanthus, and wood waste. Knowledgeable observers do not expect that this research will have a major impact on ethanol production for at least 12 to 16 years. High levels of capital are required for new biofuel processing facilities ($350 million plus per plant), and it takes about 4 years to develop pilot factories and another 8 years to move to large-scale commercialization (Sommerville, 2009). Another set of constraints involve sorting out logistics issues (e.g., harvesting, transportation, and storage of feedstocks) that will require major investments in R&D (Richard, 2010).

Also a substantial amount of private-sector (but little public-sector) money is being invested in R&D to reduce the costs of first-generation biofuels, particularly sugarcane but also corn, soybeans, rapeseed, palm oil, and others. The investment amount is considerably less than that being spent on second-generation biofuels. It is the first-generation biofuel R&D, however, that will lead to reductions in production costs for biofuels currently in use. Low-cost first-generation technology will be necessary to fulfill most U.S. biofuel needs for the next two decades until second-generation technology becomes commercially viable.

Finally, a number of small startup companies, as well as some major oil companies, are investing in algal research. These investments have grown rapidly since 2000. At present, algal research appears to be receiving less R&D funding than cellulosic research but more than first-generation research.

Three major factors are driving biofuel research: (1) the economics of biofuel—the prices of biofuels, petroleum, corn, vegetable oil, and sugar— and the cost of biomass, labor, and capital; (2) environmental concerns, which have resulted in a large number of mandates, subsidies, and regulations based on life cycle analyses (LCAs) of greenhouse gas (GHG) impacts; and (3) political and strategic issues related to national energy security.

Throughout 2010, the average monthly price of crude oil fluctuated between $75 and $85 per barrel, which is high by historical standards—but so was the price of most food commodities that could be turned into biofuel. High (relative) food prices reduce the incentive to produce biofuel from these commodities. Also, results of some recent LCAs suggest that corn and some vegetables oils may not prevent as much GHG emissions as previously thought (OECD, 2008). These factors have increased interest in second- and third-generation biofuels that are based on nonfood commodities. The conversion of biomass to biofuels, however, is still expensive and unproven commercially. Hence, interest is high in R&D to raise the productivity and reduce the cost of second-generation and algal biofuels. Because any of the biomass crops will have to be grown in areas where the cost of production is low, they will likely be grown outside of major oil-exporting countries, which makes biofuels politically attractive as a means of enhancing national energy security.

One set of policies that could affect the demand for biofuel and, thus, research investments by the private sector, are policies to slow climate change. The expectation that some form of climate change policy will be enacted that will raise the price of fossil fuels is already a major impetus for private-sector investments in biofuels R&D. If such policies are enacted, and if the current type of LCA (which finds relatively few GHG savings from first-generation biofuels) becomes the accepted measure of GHG emissions, demand for second-generation biofuels and private biofuel R&D may increase even further.

References

ACGT. 2010. Company website, available at: www.acgt.asia (accessed November 15, 2010).

Archer Daniels Midland (ADM). 2008. *2008 Annual Report*, available at: www.adm.com/en-US/investors/

BASF. 2009. CTC – "Centro de Tecnologia Canavieira – and BASF Enter Technical Cooperation Agreement in Sugarcane," News Release, August 4, 67056 Ludwigshafen, Germany, available at: www.basf.com/group/ pressrelease/P-09-147.

Biofuels Digest. 2009a. "Brazil Records Record Ethanol Exports in 2008: Report," January 27, available at: www.biofuelsdigest.com

Biofuels Digest. 2009b. "VC Investment in U.S. Biofuels Reaches $680.2 Million in 2008," *Biofuels Digest Special Report: A Summary of US Venture Capital Investment in Biofuels*, January 23.

Biomass Research and Development Initiative (BRDI). 2007. *Solicitations for the Fiscal Years 2002-2006*, U.S. Department of Energy, available at: www.brdisolutions. com/default.aspx.

BNDES & CGEE. 2008. *Sugarcane-Based Bioethanol: Energy for Sustainable Development*, BNDES, Rio de Janeiro, Brazil.

BP. 2008. "BP Announced Today That It Intends To Take a 50 Percent Stake in Tropical BioEnergia SA," BP News Release, April 24, London, available at: www.bp.com/generic article.do?categoryId=2012968&conte ntId=7043976.

Chang, Helen Kaiao. 2009. "ExxonMobil Investment Lifts San Diego's Algae Biofuel Industry," San Diego News Network July 14, www.sdnn. com/sandiego/2009-07-14/business-real-estate/mobilexxon-investmentlifts-san-diego%E2%80%99s-algae-biofuel-industry.

Davis, Ann, and Russel Gold. 2009. "U.S. Biofuel Boom Running on Empty, *The Wall St. Journal*, August 27.

F.O. Licht. 2008. *World Ethanol and Biofuels Report* 7, 5 (November). F.O. Licht. 2009. *World Ethanol and Biofuels Report* 6, 15 (April 9).

Howell, Katie. 2009. "Exxon Sinks $600M into Algae-Based Biofuels in Major Strategy Shift," *The New York Times*, July 14.

International Energy Administration (IEA). R&D Statistics, IEA/OECD, Paris, France, available at: http://data.iea.org/ (accessed November 30, 2010).

Krauss, Clifford. 2009. "Big Oil Warms to Ethanol and Biofuel Companies," *The New York Times*, May 27.

Matsuoka, Sizuo, Jesus Ferro, and Paulo Arruda. 2009. "The Brazilian Experience of Sugarcane Ethanol Industry," *In Vitro Cellular & Developmental Biology - Plant* 45:372-381.

Monsanto. 2008. "Monsanto Company To Invest in Technologies for Sugarcane With Acquisitions of CanaVialis and Alellyx," Monsanto News Releases, November 3, available at: http://monsanto.mediaroom.com/ index.php?s=43&item=656

Neste Oil. 2010. *Neste Oil Annual Report 2009*, available at: www.nesteoil. com (accessed November 15, 2010).

Organisation for Economic Co-operation and Development (OECD). 2008. *Biofuel Support Policies: An Economic Assessment.*

Praj Industries Ltd. 2011. Annual Report 2010-2011. http://www.praj.net/ media/Praj-AR-2010-11.pdf. Downloaded from the web October 11, 2011.

Rajagopal, Deepak, Steve Sexton, Gal Hochman, and David Zilberman. 2009. "Recent Developments in Renewable Technologies: R&D Investment in Advanced Biofuels," *Annual Review of Resource Economics* 1:621-644.

Richard, Tom L. 2010. "Challenges in Scaling Up Biofuels Infrastructure," *Science* 329, 793-795.

Royal Dutch Shell. 2008. "Shell Announces Six New Biofuels Research Agreements," Royal Dutch Shell News Release, September 17, The Hague, Netherlands, available at: www.shell.com/home/content/media/ news_and_library/press_releases/2008/biofuels_research_agreements_17092008.html.

Shafer, Steven. 2011. Deputy Administrator, U.S. Department of Agriculture, Agricultural Research Service, personal communication.

Sheehan, J., and M. Himmel. 1999. "Enzymes, Energy and the Environment: A Strategic Perspective on the U.S. Department of Energy's Research and Development Activities for Bioethanol," *Biotechnology Progress* 15, 5: 817-827.

Somerville, Chris. 2009. "Development of Cellulosic Biofuels," Presentation at the Conference on Biofuels in Asia, Africa and Latin America, University of California, Berkeley, May 8-9.

Sticklen, Mariam B. 2007. "Feedstock Crop Genetic Engineering for Alcohol Fuels," *Crop Science* 47:2238-2248.

Syngenta. 2007. "Syngenta Starts Research Partnership in Australia for Sugarcane Biomass Conversion to Biofuels," Syngenta News Release, October 22, Basel, Switzerland, available at: www.syngenta.com/en/ media/mediareleases/en_071022.html

U.S. Department of Energy (DOE). 2008. "DOE Announces Funding Opportunity of up to $200 Million for Pilot and Demonstration Scale Biorefinery Projects," News Release, October-November.

U.S. Department of Energy (DOE). 2009. "DOE and USDA Select Projects for More Than $24 Million in Biomass Research and Development Grants," News Release, November 12.

Weyerhaeuser. 2009a. *Company 2009 Factbook.*

Weyerhaeuser. 2009b. *2009 Annual Report and Form 10-K.*

End Notes for Research Investments and Market Structure in the Food Processing, Agricultural Input, and Biofuel Industries Worldwide: Synthesis of Results

[1] Beintema and Stads (2008) actually report fi gures in constant 2005 dollars, which we convert to current 2000 dollars using the U.S. implicit Gross Domestic Product (GDP) price index. Global totals in U.S. dollars are calculated using market exchange rates, while totals in purchasing-powerparity (PPP) dollars are derived using the PPP exchange rates. PPP exchange rates are estimated by the World Bank by comparing the cost of a common basket of consumer goods across countries. The main effect of using PPP exchange rates is to augment estimates of research and development (R&D) spending in developing countries; aggregate spending by high-income countries remains about the same whether market or PPP exchange rates are used.

[2] This breakdown of U.S. public agricultural research expenditures is according to Research Problem Areas as defi ned by USDA's Inventory of Agricultural Research (USDA, 2000). Alston et al. (2010) use a more detailed, project-by-project assignment to estimate (R&D) expenditures related to production agriculture. Their estimates show that the share of U.S. public agricultural (R&D) allocated to production agriculture has gradually declined over time.

[3] Biofuel feedstocks are the crops and biomass materials used to produce ethanol and biodiesel. First-generation feedstocks include corn, sugarcane, soybeans, and palm oil. Secondgeneration feedstocks (under development) include sources of cellulosic biomass, such as switchgrass, miscanthus, corn stover, sugar bagasse, and forestbased materials. Third-generation biofuel feedstocks include algae and synthetic life forms (see chapter 10).

[4] This is only an approximate measure of actual (R&D) expenditures within a region or country, as it includes (R&D) conducted by those same companies in other regions or countries and excludes (R&D) in those areas by companies based outside the region or country. For example, to the extent that U.S.-based companies conduct some of their R&D in foreign countries, the estimates will overstate research in the United States. But they also understate research in the United States because they exclude research conducted by foreign companies in the United States. Our assessment is that these measures are roughly correct for OECD countries, although they may understate R&D taking place in developing countries. While private-sector agricultural R&D in most developing countries is relatively small, the contribution of foreign firms to that R&D may be signifi cant. In a survey of private business enterprises in seven developing countries in Asia, Pray and Fuglie (2002) find that about 45 percent of total private agricultural R&D in those countries was conducted by foreign fi rms.

[5] Global average prices of agricultural inputs are not available, although they can be derived from trade statistics. Using trade data, we constructed global price series for farm machinery, fertilizer, and animal feed and compared these with global indexes of agricultural commodity prices. We found similar trends to the price trends for the United States shown in fi gure 1.3.

[6] The Herfi ndahl index (or Herfi ndahl-Hirschman index, or HHI) is a commonly used measure of market concentration. Higher levels of HHI indicate that sales are concentrated among a smaller group of fi rms and the potential for an increase in market power by the largest fi rms. The Herfi ndahl index is calculated as $HHI = \sum_i^N S_i^2$, where Si is the market share of fi rm i in a market with N firms. The (four- and eight-firm) concentration ratio measures the market share of the (four and eight) largest firms. Unlike the concentration ratios, the Herfi ndahl index refl ects the distribution of the market shares among the top fi rms and the composition of the market outside the top fi rms. It also gives proportionally greater weight to the market shares of the larger firms (Scherer and Ross, 1990). Note that the concentration measures in table 1.7 refer to an entire global agricultural input sector. Market concentration in a particular country or for a particular product (corn seed, or a class of herbicide, for example) could be considerably higher.

[7] These fi gures are presented to characterize the scale of private R&D, but it should not be inferred that the public and private sectors engage in similar kinds of research. Rather, each sector is likely to play complementary roles. A detailed 1994 survey of public and private crop breeding in the United States, for example, found that about 80 percent of private-sector crop breeding research was on varietal development, while breeders at USDA's Agricultural Research Service focused exclusively on more "upstream" (basic) research like developing new breeding methods and introducing new genetic diversity into breeding pools (Frey, 1996). See Fuglie et al. (1996) for more information on the roles of the public and private sectors in agricultural research and development.

[8] Bayer reports $907 million in agricultural R&D by its CropScience division in 2009, while its Consumer Health division likely spent an additional $110 million on animal health R&D (Bayer, 2010). Monsanto reports total R&D spending of $1.1 billion in 2009 (Monsanto, 2010), while Syngenta reports $960 million in agricultural R&D in the same year (Syngenta, 2010).

[9] This is not a precise estimate for developing countries, however, because the Asia-Pacifi c region includes Japan, South Korea, Australia, and New Zealand (high-income countries) with the Europe-Middle East-Africa region includes some developing countries.

End Notes for Private Research and Development for Crop Genetic Improvement

[1] By "seed," we refer to all planting material used in crop production, including seed grains, cuttings, seedlings, and other plant propagation materials. Our defi nition of the seed "market" also includes transactions for seed traits, including licensing of genetic material used in seed production.

[2] We used Context Network estimates of the value of farmer-saved seed between 2001 and 2006 and extrapolated for earlier and later years. We followed this procedure to maintain consistency in sources for the different categories of seed. However, given substantial areas planted to farmer-saved seed in developing countries, and even seed for some open-pollinated crops in highincome countries, underestimates are likely. More accurate estimates could be found by performing a country-bycountry analysis of seed markets for major crops, an exercise beyond the scope of this study.

[3] Both the estimate of regional seed market shares for Asia-Pacifi c and other estimates, like the share of the global seed market held by rice seed, are particularly infl uenced by how information from China's governmentowned seed enterprises is defi ned and measured.

[4] Although the terms of licensing and royalty agreements are, in general, not public information, most such agreements include statements that define property rights pertaining to each party to the agreement, the cost of the license, the mechanisms for royalty payments, potential philanthropic or humanitarian use, technology stewardship, and enforcement and litigation provisions (Cahoon, 2007). One feature of such agreements that has gained considerable recent attention concerns the rights of the licensee to combine the licensed technology with other technologies, either self-developed or licensed from alternative sources.

[5] Available at www.nass.usda.gov/Publications/Ag_Statistics/index.asp.

[6] Available at www.nass.usda.gov/QuickStats/Create_Federal_All.jsp.

[7] Available at www.ers.usda.gov/data/biotechcrops/.

[8] Available at www.ers.usda.gov/data/arms/app/default.aspx?survey_abb=CROP. We interpolated to calculate seeding rates for years in which they were not estimated.

[9] This method may slightly overstate the germplasm value for self-pollinating crops like soybeans, for which farmers may use part of their own harvest as seed for the next season. However, in recent years, nearly all soybean seed in the United States has been purchased. Cotton is also selfpollinating, but almost universal use of purchased seed in cotton began much earlier than it did in soybeans.

[10] Kalaitzandonakes et al. (2010) also analyze the pricing of different types of GM traits, either individually or in combination. Much of the data analyzed by Shi et al. (2009, 2010a, 2010b) and by Kalaitzandonakes et al. suggest that component pricing is generally not additive—that is, the price premium for seed containing two or more GM traits is not the sum of the individual premium for the traits taken separately.

[11] For the most part, the data presented by Shi et al. (2009; 2010a; 2010b) indicate slightly higher estimates of the proportion of crop land planted to biotech crops, somewhat higher estimates of the proportion of crop land planted to stacked varieties, and somewhat lower prices for biotech crops than the NASS data.

[12] BASF took a partial interest in Svalöf Weibull, a Swedish seed company, in 1998 but divested in 2008. Svalöf Weibull changed its name to Lantmännen SW Seed in February 2010.

[13] For example, between 2005 and 2010, Dow acquired or reached commercial agreements with nearly 10 regional seed companies, both in the United States and elsewhere, whose primary focus is corn (maize).

[14] Dupont took an initial fi nancial interest in Pioneer in 1997 and took full ownership in 1999.

[15] Estimates of seed market shares for the United States are generally not publically available, although a private source is GfK Kynetec (formerly dmrkynetec). Using GfK Kynetec data, an online Monsanto report traces the corn, soybean, and cotton seed market shares of the largest integrated seed-chemical companies between 1997 and 2009 (see www.monsanto. com/newsviews/Pages/monsanto-submission-doj.aspx). Another consulting

company, Context Network, also produces seed market share estimates, and one of its online reports estimates corn seed market shares for leading companies for 2008 (see www.contextnet.com/Focus%20Papers/Seed/Consolidation%20Direction%20Where%20and%20Why%20the%20Seed%20Industry%20is%20Headed%20Sieker%204%2008.pdf).

16 Generic fi rms are fi rms that do not conduct research on new trait discovery but rather focus on commercializing varieties and traits that have gone off-patent. So far, generic fi rms are virtually nonexistent in the crop seedbiotechnology industry, but they play an important role in the agricultural chemical industry (see chapter 3).

17 We refer to these companies as "dedicated" agricultural biotechnology companies in that agricultural applications appear to be the main market for their product development efforts. Not included in this category are many other companies that may provide some technology services to agriculture but whose main product focus is in other economic sectors. Names of these companies are available from the authors upon request.

18 Marker-assisted breeding is an indirect process in which selection for a trait of interest is based on a genetic or other marker associated with the trait, but not the trait itself.

19 In table 2.3, we include research expenditures by legacy companies that also conducted both agricultural chemical and seed-biotechnology research as part of the Big 6. These include AgrEvo, Astra-Zeneca, Aventis, Ciba-Geigy, Novartis, and Sandoz.

20 Le Buanec's estimates of biotechnology research as a percentage of total seed-biotech research in 2003 are as follows: Monsanto 80 percent; Dupont/Pioneer 50 percent; Syngenta 60 percent; BASF 100 percent; Dow 85 percent; Bayer Crop Science 85 percent; and Limagrain 10 percent. Applied to our estimates of research expenditures in 2003, this implies an estimate that about 70 percent of all Big 6 seed research was devoted to biotechnology in 2003, with the rest of our estimates based on the assumptions listed in the text. The overall estimate of 50 percent may be roughly the same today, or even higher, as large seed companies outside of the Big 6 may be increasing biotechnology investments. However, Syngenta has apparently decreased its spending on crop biotechnology while increasing its overall seed research in recent years, if the reported research expenditure devoted to "business development" in its annual reports can be assumed to be biotechnology research. This has been our working assumption for Syngenta. Other companies do not make this distinction in reporting their R&D expenditures.

21 As mentioned in footnote 17, some privately collected data are available for purchase from such companies as GfK Kynetec. These data focus particularly on large markets such as the United States, on commercial seed sales as reported by farmers, and on major crops such as corn, soybeans, and cotton for which many varieties include GM traits. Magnier et al (2010), for example, use data purchased from GfK Kynetec to develop and analyze a fairly complete list of corn hybrids sold in the United States between 1997 and 2009. These data are also the basis for Shi et al.'s (2009; 2010a; and 2010b) analysis of biotech trait pricing for U.S. corn, soybeans, and cotton.

22 Patent counts have often been used as an indicator of innovative activity. Huffman and Evenson (2006) use U.S. data on patents issued to private inventors in four agricultural areas as a proxy for private agricultural R&D capital. Brennan et al. (2005) present annual concentration ratios for both agricultural biotechnology patents and agricultural biotechnology field trials in the United States as possible indicators of competitiveness in markets for agricultural biotechnology. Nonetheless, patent data used in this way suffer from some limitations: the propensity to use patents or other forms of intellectual property protection might vary over time, by crop, and over jurisdiction (e.g., the application of utility patents to crop cultivars has been primarily a U.S. phenomenon). The propensity to patent also varies depending on the technology. Research tools, traits for genetic engineering, or GM cultivars, for example, might be more likely to be patented than nonbiotechnology research outputs. Also, intellectual property strategies may vary among companies.

23 Some of the indicators when traced across time show defi nite patterns. For example, concentration in fi eld trial permits has been consistently high (CR 4s of 60 to 100 percent) (Fernandez-Cornejo, 2004). Concentration ratios (CR 10) for broadly defi ned agricultural biotechnology patents fell over the 1990s if mergers and acquisitions are not considered. If mergers and acquisitions are accounted for, the CR 10 began to rise after 1995 (King and Heisey, 2003).

24 A GM "event" is an instance where a specifi c gene has been introduced into a particular crop. Subsequent introduction of the same gene into another variety of the same crop is not considered to be a separate event.

25 If GM event approvals are counted only once (eliminating double counting of the same GM seed approved for release in more than one country), Monsanto's share of total global approvals falls from 49 to 38 percent, since that Monsanto has been particularly active in obtaining multiple approvals for the same GM events. The two

most widely approved GM events, GTS 40-3-2 (glyphosate-tolerant or "Roundup Ready" soybeans) and MON 810 (Bt corn expressing the Cry1Ab protein) both belong to Monsanto.

[26] In addition, event approvals are sometimes issued jointly to more than one company.

[27] See www.ipm.iastate.edu/ipm/icm/node/946/print.

[28] A number of different Bt genes are available, and they are now aimed at the control of different pests (e.g., the European corn borer and rootworm), contributing to the complexity of the situation with respect to insect resistance.

End Notes for Private Research and Development for Crop Protection Chemicals

[1] In the text, we use the terms "crop protection chemicals" and "agricultural chemicals" interchangeably to refer to synthetic chemical pesticides used in agriculture. "Biopesticides" are also produced by the crop protection industry using naturally occurring biological rather than synthetic chemical substances.

[2] James (2010, appendix 2) estimates that crop protection derived from biotechnology contributed an additional $10.6 billion sales in 2009, but this amount includes the value of the crop protection chemicals bundled with the value of the seed itself. In this report, we discuss the seed market and private R&D in crop breeding in chapter 2.

[3] Some of these fi rms are owned by public research institutes and were established to commercialize new chemical discoveries by government or university research laboratories. This model is especially prevalent in China (Bryant, 2007).

[4] Chapter 2 of this report illustrates the detailed evolution of additional companies.

[5] Monsanto also represents the interaction between the chemicals and pharmaceuticals industries, which is discussed in chapter 2 in more detail. See also fi gure 2.2 for a schematic history of Bayer, Dow, BASF, and Monsanto.

[6] The physical volume of global pesticide production measured by weight of active ingredients declined between 1994 and 2001, only recovering to earlier levels of production by 2007 (EPA, 1997-2011).

[7] More than 5,700 U.S. active ingredient registrations have been cancelled out of 10,178 fi nal re-registration eligibility decisions (EPA, 2009).

[8] The classifi cation A01N includes "preservation of bodies of humans or animals or plants or parts thereof; biocides (e.g., as disinfectants, as pesticides, or as herbicides); pest repellants or attractants; plant growth regulators." It is the primary international patent classifi cation for pesticides, although it contains nonpesticide patents and probably excludes some patents that are classifi ed elsewhere but are nonetheless important for crop protection.

[9] The 1995 estimate is from EPA (1997), and the 2006 estimate is from AGROW (2007b).

[10] This analysis is based on data from the EPA Pesticide Product Information System (EPA).

End Notes for Private Research and Development for Synthetic Fertilizers

[1] Nitrogen, phosphate, and potassium are classifi ed as primary macronutrients for agriculture. Other "secondary" macronutrients are calcium, magnesium, and sulfur, which are often supplied through liming or manuring. Many micronutrients (trace elements) are also required for plant growth. These may also be applied as chemical fertilizers but are usually naturally available in soil in suffi cient quantities. In this chapter, we only consider synthetic (manufactured) fertilizer and not organic fertilizer, such as animal manure.

[2] Values of the global fertilizer market are derived by multiplying Food and Agriculture Orgacalculated on a dollar per metric ton of nutrient basis. The reference prices are for Nitrogen Ukraine Urea (44-46% N), for Phosphate U.S. Gulf Port Superphosphate (45 percent P2 O5), and for potassium Canadian Potash (60 percent K2 O).

[3] Mosaic was formed in 2004 when IMC Global and Cargill agreed to combine their fertilizer businesses. Mosaic is the largest fertilizer company in the United States and also produces potash in Michigan and New Mexico.

[4] Yara International, a Norwegian fi rm, was formed in 2003 when Norsk Hydro decided to spin off its fertilizer business as a separate company. In 2007, it acquired the Finnish fertilizer manufacturer Kemira GrowHow Oyj. Yara is the largest manufacturer of nitrogen fertilizer in the world.

[5] In nutrient-equivalent units, average fertilizer prices over 2000-2005 (in constant 2006 dollars) were $328 per metric ton of N, $383 per ton of P2 O5 , and $231 per ton of K2 0. These prices are based on the market prices reported by IMF (2009) for Urea (Ukraine), Superphosphate (U.S. Gulf Ports) and Potash (Canadian) by Haver Analytics. Prices are adjusted for infl ation using the U.S. Producer Price Index for Finished Goods (Federal Reserve Bank of St. Louis).

End Note for Farm Machinery Research and Development by Private Industry

[1] The National Science Foundation's (NSF) survey of industry R&D included a question on R&D by "product fi eld," in which equipment manufacturers would indicate their expenditures on farm machinery R&D, if any. The NSF discontinued the "product fi eld" question after 1997.

End Notes for Private Research and Development for Animal Health

[1] In this report, we use estimates of global market sales of animal health products from Vetnosis (an industry consulting fi rm; formerly, Wood Mackenzie) as reported in the International Federation for Animal Health annual reports. Vetnosis reported global market sales of $16.1 billion in 2006. PhillipsMcDougall, another industry source, estimated this global market to be $17.5 billion, while Animal Pharm placed this estimate at $18.6 billion in 2006. While industry sources vary somewhat in their estimates of total global sales for animal health products, the estimates all show similar trends in sales over time and similar market shares for different product types and animal species.

[2] We include medicinal feed additives but do not include nutritional feed additives, such as vitamins, in our animal health total. Nutritional feed additives are covered in the chapter on animal nutrition. We estimate the value of the global market for nutritional feed additives to be $4.4 billion in 2006.

[3] We have been able to put together time series estimates for sales fi gures for both first- and second-tier animal health companies and their legacy companies. Combining these data with estimates of total global sales, we can derive an estimate of total sales figures for third-tier companies by subtracting fi rst- and second-tier company sales from the global total.

[4] Horizontal mergers take place between fi rms producing similar goods or services. They may be contrasted with vertical mergers, which take place between fi rms at different points in a production process, for example when a large fi rm buys out fi rms that were formerly its suppliers.

[5] Danzon et al. (2007) examine a number of hypotheses regarding merger and acquisition activity in the pharmaceutical industry. They fi nd that among larger fi rms, mergers are a response to excess capacity arising from patent expirations and gaps in a fi rm's product pipeline. For small fi rms, mergers are primarily an exit strategy in response to fi nancial trouble. The study does not fi nd economies of scale in R&D to be a signifi cant factor in explaining mergers.

End Notes for Animal Genetic Improvement by the Private Sector

[1] A small component of the poultry genetics industry breeds ducks and specialty birds, such as colored chickens, which we do not discuss here although fi gures from this sector are included in our estimates.

[2] One other broiler breeding company, Heritage Farms (owned by Perdue), has a signifi cant presence in the United States but does not sell breeding stock in other countries.

[3] The EW Group operates its layer breeding program as two distinct companies, Lohmann Tierzucht, based in Germany, and Hy-Line International, based in the United States.

[4] The National Swine Registry was formed in 1994 as a consolidation of breed registries for four major breeds.

[5] The terms "breeder" and "seedstock producer" are more or less interchangeable, although the former tends to be used more in the dairy industry and the latter in the beef industry.

[6] In recent years, there has been renewed interest in dairy cross-breeding, both for the potential of hybrid vigor and for the incorporation of traits that may contribute to total lifetime profi tability, complementing the productivity traits of dairy animals in current intensive dairy systems. Cross-breeding has also been a relatively more important part of dairy improvement in developing countries, particularly in tropical or subtropical environments where pure European breeds might not be suited to the production environment.

[7] The potential number of progeny each year from a single bull through A.I. is much greater than the potential number of progeny from a boar using the same technique (Estienne, 1993).

[8] Embryo transfer technology can allow the production of a larger number of potential A.I. sires from elite females or "bull dams," which make it particularly useful to the A.I. companies. It can also allow greater selection intensity in females within dairy herds, although it is only profi table if the very top cows are used. With the exception of Alta Genetics, few North American A.I. companies purchased and maintained females for multiple ovulation embryo transfer (MOET) programs, working instead with cooperator breeders. More European companies (e.g., CRV, formerly Holland Genetics) did maintain their own females. Embryo transfers are available from many A.I. companies, from specialized embryo transfer businesses, and from veterinary practitioners; as with breeders of live breeding stock, these embryo transfer businesses are usually quite small scale in terms of revenues and numbers of employees (Funk, 2006; Joint Research Council, 2007).

[9] Other biotechnology companies and research institutions have been working on similar technology using genomic information, but at present, the Illumina chip provides more information at lower cost (Strauss, 2010).

[10] Alta Genetics was publicly traded from 1993 through 1999, when it was purchased by Koepon Holdings.

[11] In addition, they illustrate withincountry consolidation. Although France has around 40 A.I. cooperatives and around 20 cattle-breeding institutions, recent consolidation has left the top 5 cooperative unions with over 60 percent market share (based on UNCEIA data on semen doses). Similarly, cooperatives in Germany have consolidated, especially after the reunifi cation of East and West Germany.

[12] Genus, the global leader in cattle and swine breeding, also acquired an aquaculture breeding interest when it purchased SyGen in 2005, but it quickly sold off this enterprise to focus on its core bovine and porcine businesses.

[13] To date, the use of genetic markers in dairy cattle breeding has been more effective in reducing or eliminating deleterious qualitative traits than in increasing quantitative production traits (Funk, 2006).

[14] Data are available for the monetary value of international trade for bovine semen and live bovine breeding animals from the UN's COMTRADE database. In recent years, the value of live breeding animals traded has been two to three times the value of semen traded. However, most of the trade in breeding animals is among geographically proximate countries. For example, a number of European countries report relatively high exports of live bovine breeding animals, but most of this trade takes place within Europe (Gollin et al., 2009; D. Gollin, personal communication).

[15] NAHMS is conducted periodically on nationally representative samples of U.S. farms producing hogs, cattle, poultry, fi sh, and other species.

[16] According to USDA's Agricultural Resource Management Survey (ARMS), which selects different commodities each year for intensive study, A.I. was used on 85.9 percent of dairy cows in 2005 and 14.5 of beef cows in 2008 (based on calculations by the authors from raw ARMS data). ARMS focuses on farms in major producing States. USDA's NAHMS survey for cattle suggests somewhat lower rates for A.I. According to NAHMS, 72.5 percent of all dairy cow pregnancies in 2007 resulted from A.I. Also, only 1.4 percent of beef cow females were inseminated artifi cially and 5.0 percent were both inseminated artifi cially and exposed to bulls. Over 90 percent of all beef cow females were exposed only to bulls, according to this study. A.I. was somewhat more prevalent for beef heifers than for beef cows (USDA, 2009a; 2009b). Differences in estimates may stem not only from differences in sampling frameworks but also from differences in defi nitions, especially for dairy.

[17] Embryo transfer quantity data are available from such sources as the International Embryo Transfer Society (www.iets.org) and the American Embryo Transfer Association (www. aeta.org).

[18] The term "seedstock" refers to young juvenile animals that are grown out in aquaculture facilities. "Broodstock" refers to sexually mature fi sh that are kept separate for breeding and seedstock multiplication purposes.

End Note for Private Research and Development for Animal Nutrition

[1] Our estimate of the average global wholesale price of compound feed is a weighted average of the price for corn (U.S. No 2 Yellow, FOB Gulf Ports) and soybean meal (Chicago soybean meal futures, fi rst contract forward, minimum 48 percent protein). The weights are 90 percent for corn and 10 percent for soymeal. We then assume a 30-percent markup over the cost of raw ingredients to cover manufacturing and distribution costs. Corn and soybean meal commodity price data are from the International Monetary Fund.

End Notes for Research and Development in the Food Manufacturing Industry

[1] The food and beverage manufacturing sector transforms raw agricultural materials into intermediate foodstuffs, animal feed, or edible products. It does not include the food wholesale, retailing, or service sectors. The term "manufacture" is used in the International Standard Industrial Classifi cation (ISIC) and North American Industry Classifi cation System (NAICS) codes. Several ERS publications refer to "processing" industries, as do Gopinath and Vasavada (1999). See also ERS briefi ng room "Food Marketing System in the U.S.: Food and Beverage Manufacturing," www.ers.usda.gov/ briefi ng/foodmarketingsystem/processing. htm.

[2] The term "turnover" is used by some companies, particularly those in the EU. It refers to net external revenue, which may be from product sales but which also may include additional sources of income (e.g., interest or royalties).

[3] OECD draws upon national offi cial sources to compile its statistics. For the United States, industry R&D spending comes from surveys conducted by the National Science Foundation and the U.S. Census Bureau.

[4] See ISIC revision 3 and revision 4. 5 Eurozone countries include Austria, Belgium, Finland, France, Germany, Greece, Ireland, Luxembourg, Italy, Netherlands, Portugal, and Spain.

End Notes for Private Research and Development for Biofuel

[1] Annual reports of Exxon, BP, ConocoPhillips, Chevron, Shell, Total, ENI, Petrobras.

[2] This is the amount CanaVialis and Allylex reported spending on R&D in 2008 just before their acquisition by Monsanto (BNDES & CGEE, 2008). Personal communication with Monsanto offi cials in Brazil in August 2010 confi rmed that this level of research has been maintained since the takeover by Monsanto.

[3] Second-generation biofuels are ethanol and biodiesel that were made from cellulosic biomass, such as corn stover, sugarcane bagasse, switchgrass, miscanthus, and forest-based material.

[4] The Organisation for Economic Cooperation and Development consists mainly of the free-market high-income nations in North America, Europe, and the Asia-Pacifi c region.

[5] This is a conservative estimate of USDA intramural research on biofuels. The fi gure of $31 million for ARS biofuels research relates specifi cally to the Bioenergy National Program and does not include research on genetic development of key crops. In addition, the USDA total of $69 million does not include biofuels-related research by USDA's Forest Service (Shafer, 2011).

In: Global Agricultural Industries
Editors: Shirley O. Kelley and Mike F. Grant

ISBN: 978-1-62257-476-6

Chapter 2

Public Agricultural Research Spending and Future U.S. Agricultural Productivity Growth: Scenarios for 2010-2050[*]

Paul Heisey, Sun Lin Wang and Keith Fuglie

Major Findings

- By 2050, global agricultural demand is projected to grow by 70-100 percent due to population growth, energy demands, and higher incomes in developing countries. Meeting this demand from existing agricultural resources will require raising global agricultural total factor productivity (TFP)[1] by a similar level. Maintaining the U.S. contribution to global food supply would also require a similar rise in U.S. agricultural TFP.
- TFP growth in U.S. agriculture is predicated on long-term investments in public agricultural research and development (R&D). Productivity growth also springs from agricultural extension, farmer education, rural infrastructure, private agricultural R&D, and technology transfers, but the force of these factors is compounded by public agricultural research.
- The rate of TFP growth (and therefore output growth) of U.S. agriculture has averaged about 1.5 percent annually over the past 50 years. Stagnant (inflation-adjusted) funding for public agricultural research since the 1980s may be causing agricultural TFP growth to slow down, although statistical analyses of productivity growth trends are inconclusive.
- ERS simulations indicate that if U.S. public agricultural R&D spending remains constant (in nominal terms) until 2050, the annual rate of agricultural TFP growth will fall to under 0.75 percent and U.S. agricultural output will increase by only 40 percent by 2050. Under this scenario, raising output beyond this level would require bringing more land, labor, capital, materials, and other resources into production.

---

[*] This is an edited, reformatted and augmented version of U.S. Department of Agriculture, Economic Research Service, Economic Brifef Number 17, dated July 2011.

- Additional public agricultural R&D spending would raise U.S. agricultural productivity and output growth. Raising R&D spending by 3.73 percent annually (offsetting the historical rate of inflation in research costs) would increase U.S. agricultural output by 73 percent by 2050. Raising R&D spending by 4.73 percent per year (1-percent annual growth in inflation-adjusted spending) would increase output by 83 percent by 2050.

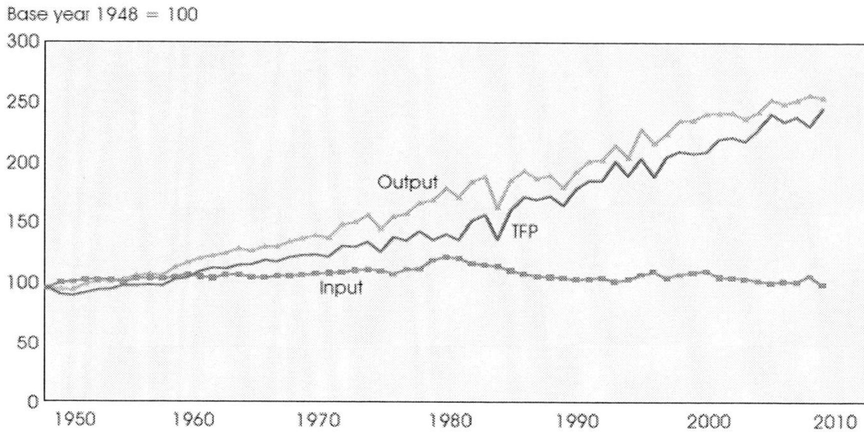

Source: U.S. Department of Agriculture, Economic Research Service.

Figure 1. U.S. agricultural output, input, and TFP indexes.

Productivity Growth Accounts for Nearly All U.S. Agricultural Growth

Total factor productivity (TFP) is a good indicator of technological change. It measures the efficiency with which all inputs (land, labor, capital, and materials) are combined to produce total outputs (all crop and livestock commodities). In U.S. agriculture, growth in TFP is nearly synonymous with growth in output since the overall size of the resource base has barely increased over the last 60 years. Between 1948 and 2008, the average annual growth rate in U.S. agricultural output was 1.58 percent and the average annual growth rate in TFP was 1.52 percent. As a result, total agricultural output in 2008 was 2.5 times that of 1948 (figure 1).

R&D Spending Spurs TFP Growth

Growth in TFP is strongly associated with the adoption of new technologies that raise yields or lower costs. Public investment in agricultural R&D is a major source of new agricultural technology; it also complements (raises returns to) other productivity-enhancing activities like extension, education, infrastructure, and private R&D. While private-sector investment in agricultural R&D also contributes new technology, much of this contribution is already accounted for in the way the inputs are measured in constructing the estimates of

TFP.[2] Moreover, studies have shown that private R&D depends critically on government and university investments in science and technology.[3] Thus, even if the quality-adjusted measures of TFP do not fully net out the contributions of private R&D, if private R&D follows (responds positively to) trends in public R&D, then simulations using public R&D as the primary policy lever for future productivity growth should not be unduly biased. Nonetheless, the simulations reported here depend on these assumptions about private R&D: that its impacts are partially netted out of the TFP measure and that it follows trends in public R&D.

Statistical models of TFP growth treat R&D spending as "knowledge capital." As with physical capital, an R&D investment can affect productivity growth for several years (even decades), but eventually depreciates as technologies grow ineffective (e.g., new pests and diseases evolve) or obsolete. R&D investments typically begin boosting TFP within 3-5 years, with benefits peaking after 10 to 20 years, and with some impacts lasting as long as 50 years (Huffman and Evenson, 2006; Alston et al., 2010).

AGRICULTURAL R&D SPENDING SCENARIOS ACCOUNT FOR INFLATION IN RESEARCH COSTS

Between 1927 and 2009 (the period for which we have data), total spending by Federal and State agricultural research institutes on productivity-related agricultural research (excluding R&D on postharvest, environmental, and rural development issues) rose at an average annual rate of 7.49 percent in nominal terms. If adjusted by the Consumer Price Index, then research spending rose by 4.37 percent per year. However, if inflation adjustments reflect the cost of doing research (i.e., scientist salaries, laboratory equipment, etc.), then real research spending rose by only 3.03 percent per year (i.e., research costs have been rising faster than the average price level of goods and services in the economy).

Estimates of R&D knowledge capital are based on real research spending (adjusted by the rate of inflation in the cost of research resources). Growth in public agricultural research spending slowed dramatically after the early 1980s and has declined by more than 20 percent since peaking in 1994 (figure 2). The growth in the stock of agricultural knowledge capital, which follows trends in research spending with a lag, peaked around 2005. Since then, knowledge capital stock has started to decline as current spending (creation of new knowledge capital) has failed to keep up with knowledge capital depreciation.

We simulate three possible scenarios for future public agricultural R&D expenditure for 2010-2050 (see Box for a description of the simulation model). These scenarios include total expenditures on productivity-oriented agricultural research by USDA research agencies and State agricultural experiment and veterinary medicine colleges from all sources of funding (USDA, other Federal and State governments, and others).

Scenario 1: Public research spending is constant in nominal terms at the 2005-09 average ($2.5 billion per year). This implies a decline in real research expenditure by 3.73 percent per year (the inflation rate in the cost of research resources over 1983-2009).[4] In other words, under this scenario, inflation reduces the effective amount of research effort.

Scenario 2: Public research spending is constant in real terms at the 2005-09 average. This implies that nominal expenditures for public agricultural research rise by about 3.73 percent per year to offset inflation in the cost of research.

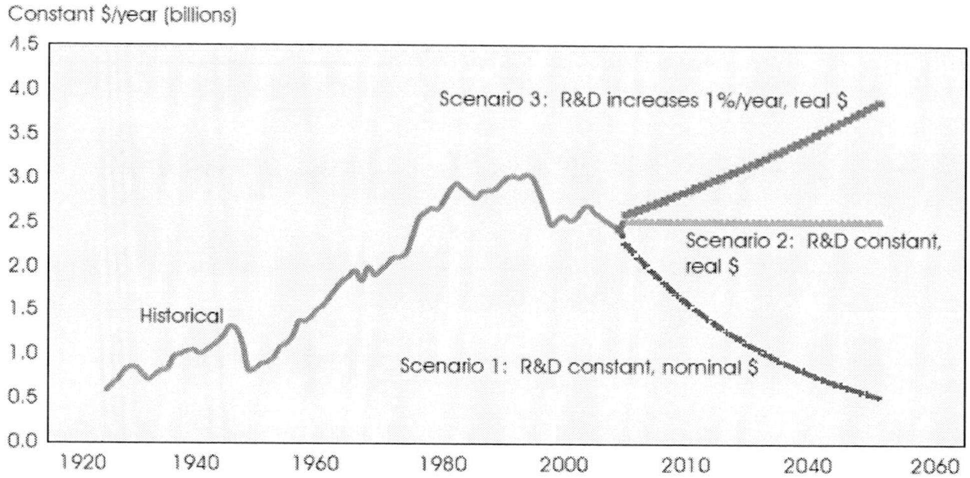

Source: Historical data from Huffman (2010) and 2010-2050 data from U.S. Department of Agriculture, Economic Research Service simulations.

Figure 2. Productivity-oriented agricultural research expenditures.

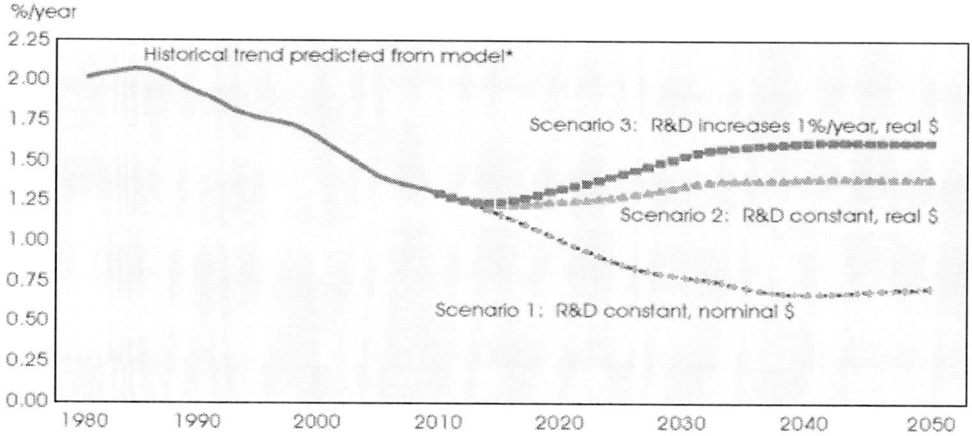

* The graph shows the TFP growth rate predicted from the statistical model rather than the actual TFP growth rate for 1980-2008. Actual TFP growth is highly variable, ranging between -15% and + 15% for individual years. This variability is mainly due to weather.

Figure 3. TFP growth rate projections.

Scenario 3: Public research spending increases by 1 percent per year in real terms (implying a nominal increase of about 4.73 percent per year). This 1-percent real growth rate is roughly equivalent to the rate of U.S. population growth (a proxy for growth in domestic agricultural demand). Real research expenditures rise, but more slowly than the rapid growth in real spending observed before the 1980s.

R&D Spending Must Rise to Maintain Historical Agricultural TFP Growth Rate

The resulting TFP projections under the three public R&D spending scenarios are shown in figures 3 and 4. Figure 3 shows the projected *annual growth rate* in agricultural TFP. The estimated historical trend based on our model shows a declining rate of TFP growth since peaking at just over 2 percent per year in the mid-1980s. By 2008, annual TFP growth was about 1.3 percent.[5] In each of the scenarios, rates of TFP growth increase or decrease from present levels and then level off after 2030 or 2040.

Under scenario 1, the TFP growth rate declines to under 0.75 percent per year. Under scenario 2, TFP growth stabilizes at about 1.4 percent per year. Under scenario 3, TFP growth accelerates and reaches 1.6 percent per year by 2040.

Figure 4 translates TFP growth into an agricultural TFP index (base year of 2008=100) under alternative funding scenarios. It also shows the agricultural output level (relative to 2008 production) achievable if agriculture's current resource base remains unchanged. Under scenario 3, by 2050 U.S. agricultural output would be 83 percent higher than 2008 levels and under scenario 2, output would grow by 73 percent. As a point of comparison, current estimates suggest global food demand will increase between 70 and 100 percent by 2050 (Food and Agriculture Organization, 2006).

Under scenario 1, however, with constant R&D funding (in nominal terms), U.S. agricultural output would increase by only about 43 percent between 2008 and 2050 using the current agricultural resource base.

This would barely outpace the projected growth in U.S. population. By contrast, U.S. agricultural output would likely keep up with the growth in global demand under the higher R&D investment scenarios.

Another recent study (Alston et al. 2010) used different data and modeling assumptions to independently derive statistical relationships between public agricultural R&D spending and TFP growth. Their findings are broadly similar to our own: public spending on agricultural R&D will need to rise to maintain historical rates of productivity growth in U.S. agriculture.

Implications of a decline in the long-term growth rate in U.S. agricultural productivity are significant for both the United States and the world. Slower productivity growth will likely cause agricultural prices to rise, which would reduce global economic welfare and raise poverty in urban areas and among food-deficit rural households, especially in developing countries (Nelson et al., 2010).

Rising prices would also increase pressure to expand agricultural cropland at home and abroad and input use on cropland would likely intensify. Such agricultural resource expansion and intensification could come with significant environmental costs, such as further impairment of soil and water quality, loss of biodiversity, and increased emissions of greenhouse gases (Foresight, 2011).

Finally, agricultural exports may also decline as U.S. farmers lose competitiveness in international markets (Ball et al., 2010).

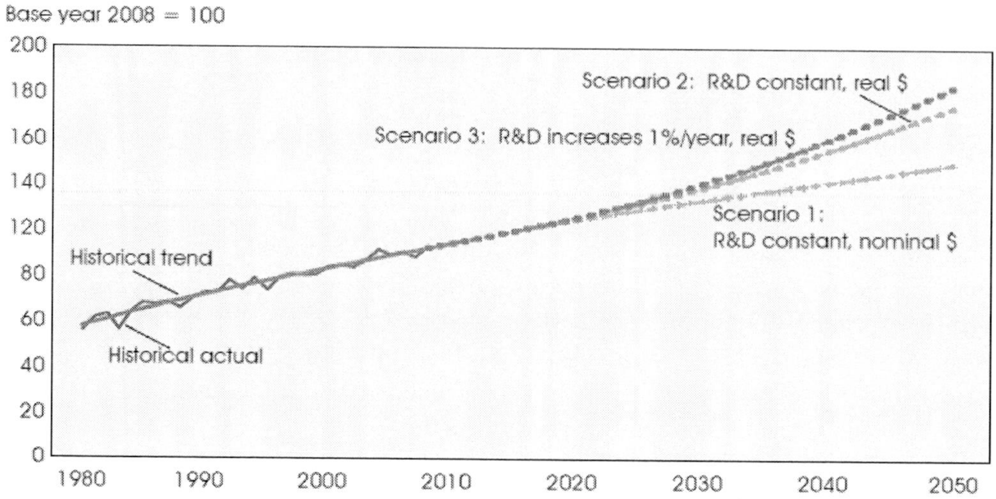

Figure 4. TFP index projections.

How the Estimates Were Derived

We use a model developed by ERS (Wang et al., 2010) along with research expenditures data from Wallace Huffman at Iowa State University. Huffman isolates the component of total public agricultural R&D associated with raising crop and livestock productivity and excludes other kinds of R&D (Huffman et al., 2001; Huffman and Evenson, 2006; Huffman, 2010).

Statistical models that link TFP to these investments treat R&D spending as "knowledge capital." Like physical capital, knowledge capital contributes to growth over a long period of time but eventually depreciates; it requires new R&D investment to be maintained and expanded.

We use the assumptions developed by Huffman and colleagues to convert annual R&D investments, or flows, into R&D capital, or stocks. Under these assumptions, R&D investments influence R&D stocks, and therefore TFP, for up to 35 years.

We assume alternative scenarios for future public R&D funding to simulate future growth in U.S. agricultural total factor productivity to 2050. We use statistical relationships based on different productivity and investment patterns across States over 1980-2004 to estimate the effects of R&D stocks on productivity growth.

These estimated relationships are used to project future TFP growth patterns given alternative assumptions about future R&D spending. Changes in current levels of public R&D spending affect future TFP growth only gradually since most of today's "knowledge capital stock" is the result of past, accumulated investment in R&D.

In order to model the effects of different scenarios for R&D spending on future TFP growth, we hold the contributions to agricultural productivity from other sources—such as agricultural extension, farmer education, infrastructure, economies of scale, and technology transfer—constant and allow only the effects from public agricultural R&D to change.

REFERENCES

Alston, Julian M., Matthew A. Andersen, Jennifer S. James, and Philip G. Pardey. 2010. Persistence Pays: U.S. Agricultural Productivity Growth and the Benefits from Public R&D Spending. Springer.

Ball, V. Eldon, Jean-Pierre Butault, Carlos San Juan, Ricardo Mora. 2010. "Productivity and international competitiveness of agriculture in the European Union and the United States." Agricultural Economics 41: 611-627.

Economic Research Service. "Data Set: Agricultural Productivity in the United States," Agricultural Research and Productivity Briefing Room. Available at http://www.ers.usda.gov/Data/AgProductivity/ (accessed June 1, 2011).

Food and Agriculture Organization. 2006. World agriculture: towards 2030/2050, Interim report. Food and Agriculture Organization of the United Nations, Rome, June. http://www.fao.org/docrep/ oo9/a0607e/ a0607e00.htm

Foresight. 2011. The Future of Food and Farming: Challenges and Choices for Global Sustainability. Final Project Report. The Government Office for Science, London, United Kingdom.

Fuglie, Keith O. and Thomas S. Walker. 2001. "Economic Incentives and Resource Allocation in U.S. Public and Private Plant Breeding. Journal of Agricultural and Applied Economics 33 (3): 459-473.

Huffman, W. E., A. McCunn, and J. Xu. 2001. "Public Agricultural Research Expenditures with an Agricultural Productivity Emphasis: Data for 48 States, 1927-1995." Iowa State University, Department of Economics, Staff Paper.

Huffman, W. E., and R.E. Evenson. 2006. Science for Agriculture—A Long-Term Perspective. Second edition, Blackwell Publishing.

Huffman, W.E. 2010. "Measuring Public Agricultural Research Capital and Its Impact on State Agricultural Productivity in the U.S." Iowa State University, Department of Economics, Mimeo. (Updates data from Huffman et al., 2001)

Nelson, Gerald C., Mark W. Rosegrant, Amanda Palazzo, Ian Gray, Christina Ingersoll, Richard Robertson, Simla Tokgoz, Tingju Zhu, Timothy B. Sulser, Claudia Ringler, Siwa Msangi and Liangzhi You. 2010. Food Security, Farming and Climate Change to 2050. International Food Policy Research Institute, Washington, DC.

Wang, Chenggang, Yin Xia, and Steven Buccola. 2009. "Public Investment and Industry Incentives in Life-Science Research." American Journal of Agricultural Economics 91(2): 374-388.

Wang, Sun Ling, Eldon Ball, Lilyan Fulginiti, and Alejandro Plastina. 2010. "Benefits of Public Research Spill-in, Extension, and Road Networks in U.S. Agriculture." (Under revise and resubmit with a peer reviewed journal).

End Notes

[1] Total factor productivity (TFP) is the broadest measure of productivity. It compares the total output of a sector to the total land, labor, capital, and material inputs used to produce that output. Increases in TFP imply more output is forthcoming from a given level of inputs, or, equivalently, fewer inputs are required to produce the same output. Growth in TFP is considered to be an indicator of the rate of technical change in a sector.

[2] New technologies from private R&D are partially embodied in the quality of the inputs (chemicals, machinery, seed, etc.) purchased by farmers. The TFP measure used in this study adjusts for such quality improvements in these inputs. In other words, it already nets out many of the productivity effects of private R&D.

[3] Public-sector research opens up new technological opportunities for the private sector to commercialize. One recent study found that "government expenditures in both basic biological research and agricultural and medical science create substantial spillovers for private firms...Indeed, opportunities created through public research are the principal source of growth in industry life sciences" (Wang, Xia and Buccola, 2009). Fuglie and Walker (2001) also found complementarity between public and private investments in crop genetic improvement, with public research in basic plant breeding spurring more private investment in crop variety development.

[4] While the average rate of inflation in research costs over 1927-2009 was 4.46 percent, this has slowed in more recent years. To project research cost inflation in the future, we chose the average inflation rate in research costs during 1983-2009, or 3.73 percent per year.

[5] This apparent decline in TFP growth is based on the model used to project future TFP growth. All other factors were kept fixed except for the changes in R&D "knowledge capital" stock; therefore, the decline in the TFP growth rate reflects the erosion of R&D stocks, as hastened by the stagnation in R&D funding that began in the 1980s (see figure 2). Thus, the slowing of TFP growth in figure 3 after the mid-1980s reflects this model assumption; actual agricultural TFP growth may not have slowed if other contributing factors offset the decline in R&D capital.

INDEX

D

E

F

G

H

I

J

K

L

M

N

O

P

Q

R

S

T